De Havilland
MOSQUITO
PORTFOLIO

Compiled by
R.M. Clarke

ISBN 0 948 207 914

Published by Brooklands Books in conjunction with

AEROPLANE MONTHLY

A BROOKLANDS AIRCRAFT PORTFOLIO

Portfolios in this series

De Havilland MOSQUITO
Hawker HURRICANE
Boeing B-17 and B-29
FORTRESS AND SUPERFORTRESS
Handley Page HALIFAX

Portfolios in preparation will cover: Spitfire, Lancaster, Wellington, P51 Mustang etc.
Cover Photography by Flight International

DISTRIBUTED BY

Motorbooks International
Osceola
Wisconsin 24020
USA

Brooklands Book Distribution Ltd
Holmerise, Sevenhills Road
Cobham, Surrey KT11 1ES
England

A BROOKLANDS AIRCRAFT PORTFOLIO

A BROOKLANDS AIRCRAFT PORTFOLIO

This is the first of a new series of books covering classic World War Two aircraft, compiled from contemporary material originally published in Flight, The Aeroplane *and* Aircraft Production. *We have combed through wartime volumes of these journals and selected representative features on these famous fighting aircraft, covering every aspect of each aeroplane: genealogies, technical appraisals, handling characteristics, combat and operational reports and so forth. In addition to a wealth of photographs each book contains one or more contemporary cutaway drawings, for which both* Flight *and* The Aeroplane *were noted. These articles have not been edited in any way and are straight, high-quality reprints from the original issues.*
In addition to this wealth of contemporary material further features have been reprinted from Aeroplane Monthly, *successor to* The Aeroplane, *written with hindsight by pilots, engineers and crewmen who knew their aircraft intimately, inside and out.*
Whether or not you flew in these aircraft, this series conveys, in a unique form, the character and background of aircraft that fought for peace during six long years of war.
If you enjoy this book you will surely enjoy others in this series.

Richard Riding
Editor
Aeroplane Monthly

DE HAVILLAND MOSQUITO

VERY little information has as yet been released for publication about the De Havilland Mosquito, the R.A.F.'s latest, and probably fastest, light reconnaissance-bomber.

This machine, as described in *Flight* of October 29th, is of wooden construction and is powered by two Rolls-Royce liquid-cooled, 12-cylinder "V" type engines, equipped with De Havilland three-bladed, hydromatic type airscrews, and although performance figures may not yet be mentioned, it goes without saying that they are extremely good. Undercarriage and tailwheel are retractable, but the latter is not completely enclosed.

It is, as the accompanying illustrations show, a particularly clean aircraft, with a beautifully streamlined fuselage on which the cockpit cover, while obviously providing good visibility, blends very smoothly into the upper surface.

Another good feature, aerodynamically, is the way in which the engine nacelles are underslung, so that their upper surfaces merge into the upper surface of the wings, an arrangement which has been found to give very low drag, as the airflow over the top of the wing suffers a minimum disturbance by the power units.

The Mosquito was first officially mentioned when four of them made a daylight raid on the Nazi headquarters at Oslo; since then the type has increasingly figured in successful raids on enemy objectives across the Channel.

Incidentally, the Mosquito is the first operational type to be developed by De Havilland since the last war, and its simple construction lends itself to widely dispersed manufacture.

Offensive armament may consist of four 20 mm. cannon and four 0.303 machine guns.

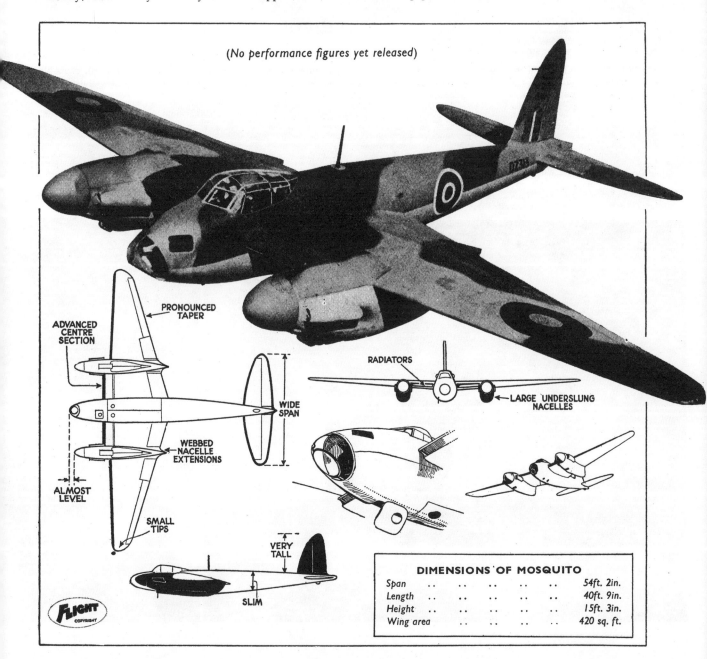

(No performance figures yet released)

PRONOUNCED TAPER

ADVANCED CENTRE SECTION

WIDE SPAN

WEBBED NACELLE EXTENSIONS

ALMOST LEVEL

SMALL TIPS

RADIATORS

LARGE UNDERSLUNG NACELLES

VERY TALL

SLIM

DIMENSIONS OF MOSQUITO	
Span	54ft. 2in.
Length	40ft. 9in.
Height	15ft. 3in.
Wing area	420 sq. ft.

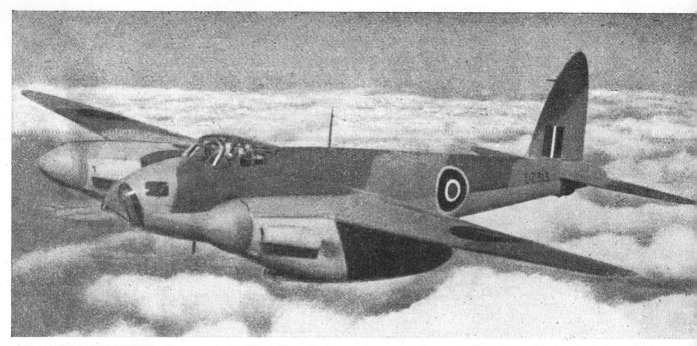

DE HAVILLAND'S LATEST.—The new de Havilland Mosquito reconnaissance bomber (two Rolls-Royce Merlin motors), probably the fastest bomber in the World. Something of the lines of the famous D.H. Comet can be seen in the Mosquito.

THE FASTEST BOMBER.—Two de Havilland Mosquito high speed reconnaissance bombers flying in formation with the photographic Boston. Powered with two Rolls-Royce Merlin 21 motors, the Mosquito is the fastest aeroplane of its type in the World to-day.

THE NEW RECONNAISSANCE BOMBER : These pictures, taken by our chief photographer, John Yoxall, give a fine impression of the manoeuvrability of the De Havilland Mosquito (two Rolls-Royce engines). All three wheels retract, leaving a particularly clean exterior, increased by the wooden construction which is without stress wrinkles. Offensive armament may comprise four 20mm. cannon and four .303 machine guns.

MOSQUITOES IN THE DAY

BOMBER COMMAND is now using de Havilland Mosquito bombers for low level attacks on targets in Western Europe. Manned by picked air crews, these bombers have been taking part during the past two months in the offensive of Fighter, Bomber, and Army Co-operation Commands, particularly against the transport and communications systems under the control of the Germans.

Only very small numbers of Mosquitoes have failed to return from operations on the other side of the English Channel and North Sea. The defensive characteristics required by Mosquitoes to protect themselves from the cannon and machine-gun fire of the Focke-Wulf and Messerschmitt fighters of the Luftwaffe lie in the aerodynamic efficiency and sound workmanship of the aeroplane, which closely follows the almost perfect aerodynamic form of the de Havilland Albatross, combined with the immense power compressed into the two Rolls-Royce Merlin 21 motors.

Military specifications must always include some items which will break the clean lines of a prototype or civil aeroplane. In the Mosquito the nose has had to be flattened to allow the Observer to use the bomb sight when operating at high levels for precision bombing. This, with one or two other imperfections, has made the performance of the operational Mosquito slightly lower than that of a possible civil mail-carrying version, but even so, the top speed of the Mosquito at "deck" level is exceptional.

The performance of the Mosquito does not end with its high speed, but also includes a very fast rate of climb and a range which is long enough to enable squadrons equipped with Mosquitoes to operate over Norway and far inside Germany itself. This range is combined with a high cruising speed well above that of any other bomb-carrying aeroplane.

When one Mosquito was attacked by a Focke-Wulf Fw 190A3 it went into so tight a turn that the Fw 190 with its wider turning radius was unable to get the bomber in its gun sight.

At low levels the performance of the Mosquito is excellent. The squadrons have, therefore, been trained in the new methods of low level, hedge hopping attacks and now the de Havilland Mosquito is being used in conjunction with the Mustangs, Spitfires and Bostons of Army Co-operation and Fighter Commands in their combined attacks on Western Europe.

Attacks are often made either at dusk, after approaching a target in daylight to return in the dark, or at dawn after setting out in the early hours of the morning to return in time for breakfast, although some raids are made in the hours of broad daylight.

One Mosquito Squadron is commanded by Wing Commander H. I. Edwards, V.C., D.S.O., D.F.C., who has 53 operational sorties to his credit, and includes Squadron Leader D. A. G. Parry, D.S.O., D.F.C., who led the raid on the Nazi Party Headquarters at Oslo, Squadron Leader J. R. G. Ralston, D.S.O., D.F.M., hero of 74 sorties, including the unique operation of blocking a railway tunnel in France at both ends while a train was inside, and Flight Lieutenant S. C. Clayton, D.F.C., D.F.M., who has been on more operational flights than any other Observer in Bomber Command (he only wants another seven to complete his century).

The crews gather together in the crew room. Aircraft recognition charts, supplemented with numbers of solid models, hung from the ceiling, provide a constant reminder of the possibilities of meeting enemy aircraft. Instructions in the form of charts cover the walls and include the procedure to be adopted when baling out. Both motors have to be stopped because the airscrew's arc approaches near to the escape and entrance hatch in the floor of the cockpit.

The charts detail dinghy procedure in case a Mosquito has to alight on the sea and give information about the Air-Sea rescue service. All members of the air crews of Mosquitoes carry a whistle attached to the right lapel of their battle dress blouses as a substitute for shouting to attract attention.

There are notes on the ill effects of cold and height on the fighting qualities of air crews, a diagram of the flare path system in use on the home aerodromes, drawings of ships both Allied and enemy, the new phonetic alphabet which has been adopted by the R.A.F. for radio telephony, information on the handling of D.H. airscrews, and a drawing of the oxygen economiser system.

Other charts of a more highly secret nature give information about the controls and coolant systems of the Mosquito, and yet another gives hints on how to obtain the best operational performance. Apparatus is provided for testing flying helmets before leaving the crew room. A lighter side which takes the pilots' minds away from the ever present reminders of war is provided by the constantly used radio set, a chess board, and numbers of topical magazines which lie on the large table in the centre of the room.

On a sortie, the two members of the Mosquito's crew are accommodated in the forward part of the fuselage. The rear part, which contains the oxygen, radio, and recognition lights, as well as a large amount of other fixed equipment, is sealed before take-off and no entry to it is possible in flight. The pilot sits on the left of the cockpit with his observer on his right. A prone position on the right-hand side of the nose is left for the use of the Observer, but it is only used for high level precision bombing. All navigation up to within seven or eight miles of a target is the duty of the Observer who also operates the two-way radio while keeping a look-out at the back of the machine for hostile fighters. When Mosquitoes operate in pairs or larger formations, all the navigation is done by the Observer in the leading aeroplane while the Observers in other machines watch for enemy aircraft.

Once aboard the air crews settle down in their places, check all the equipment, rev. up the motors, close the bomb doors, operated by a hydraulic system (as is the undercarriage), on four 500-lb. bombs and prepare to take off.

Then with main and tail wheels retracted the Mosquitoes swiftly climb and streak across the English countryside. Over the sea they go down to a slightly lower level before making a landfall. This landfall is said to be one of the most important parts of a sortie. If an aeroplane can fly straight over the coast up to its target it will strike with all the advantages of surprise on its side. But if an aeroplane flies on to its target after flying up and down the coast trying to find the correct landmark, it will arrive to find a hot reception prepared for it. As only one bombing run is made, landmarks are followed right up to the target. Five or six miles before it, the pilot takes over the navigation, opens the bomb doors, and advances the throttle. Then, going in at "nought" feet, the Mosquitoes flash over their targets dropping all their bombs. These bombs are fitted with delayed-action fuses to enable the aeroplane to get anything up to a mile away before they explode.

When marshalling yards are being attacked the bombs often hit hard sur-

EXPRESS DELIVERY.—A de Havilland Mosquito in the background awaits its load of four 500-lb. bombs to be carried in daylight to Germany at a higher speed than in any other bomber in the World.

OFFENSIVE

faces and because they are dropped so low at high speed, they strike with glancing blows. As a result they often bounce or ricochet sometimes even higher than the aeroplanes which dropped them, to go off eventually in some place where they will succeed in helping to cripple the German War effort.

On the run up to the target, Mosquitoes may attack in formation. After bombing, the machines break formation and race for home.

Two typical dusk attacks took place on January 3, when small formations of Mosquitoes attacked marshalling yards in Northern France. They gave an opportunity for the Royal Air Force to demonstrate the destructive power of its light day bomber squadrons.

Thorough preparations were made before the raid. The crews learnt by heart the contours of the ground around their targets and in the late afternoon six Mosquitoes took off to raid marshalling yards. When a mile away from the target, light anti-aircraft fire engaged the raiders, flying in echelon to starboard with their bomb doors open. None of the machines was hit, and they still escaped undamaged when some heavy A.A. batteries opened up with short-fused shells. These burst above the aeroplanes and were completely ineffective.

Running straight on to the target, all our aircraft dropped their bombs and wrecked rolling-stock, the rolling-stock repair shop, and engine sheds. After closing their bomb doors the Mosquitoes returned without loss, in formation, to their aerodromes somewhere in England.

While this raid was taking place more Mosquitoes were attacking another marshalling yard. After making a successful landfall on the French coast, the aeroplanes flew across country low enough to let one pilot bring back to this country some of France's brushwood. All the bombs landed in the target area and one of the pilots insists that no trains left the target area that night.

One pilot recently had a running fight with a Focke-Wulf Fw 190 which lasted for more than 20 minutes. At first the Mosquito tried to shake off the Fw 190 by going round in tight turns, but his opponent followed. Seeing that the position was not at all favourable at the height at which they were flying, the crew of the Mosquito brought their machine down to ground level, taking evasive action all the time. Low down the Mosquito was able to show its heels to the Fw 190, which was mainly concerned with avoiding the trees which persisted in getting in the way. The German pilot was so concerned with the trees that he let the Mosquito escape after having fired only two bursts, both of which went wide.

On another occasion the same Mosquito was attacked by a Messerschmitt Me 110. This time the speed and climb of the British aeroplane enabled it to slip up into a convenient cloud, completely baffling the slower Messerschmitt.

During the hours of daylight pilots have observed many of the population in Holland and the other occupied territories standing in the open waving to our aircraft. On the other hand, Germans usually dive for cover at the approach of any British aeroplanes. At night the friendly attitude of the Dutch population can be judged from the numbers of " Vs " which are incessantly flashed into the air from torches and windows. One of our pilots flew across Holland one night flashing the letter V from his recognition lamp and he found the sign flashed back all along his route.

From conversations with Sqdn. Ldr. Parry and other members of the air crews who took part in the raids on Oslo, we gather that the Nazis on that day had flown many of their latest aeroplanes to Kjeller aerodrome, near Oslo, to present a flying exhibition, which they hoped would bolster up the moral of their troops and impress the Norwegian population. When our four Mosquitoes arrived over Oslo, two Fw 190s were already in the air. Our pilots were extremely surprised to find such aeroplanes there but even more surprised were the Fw 190 pilots to find such high-speed bombers operating over Norway.

DIVE APPROACH.—The graceful lines and fine streamline of one of the World's best aeroplanes is seen in this photograph of the D.H. Mosquito (two Rolls-Royce Merlin 21 motors.)

The Mosquitoes were flying in line astern and both Fw 190s attacked the rear machine, which was last seen going down with clouds of thick black smoke coming from one motor. Nothing was seen of the Fw 190s after that for a second or two until the Observer in the third Mosquito saw cannon and machine-gun fire going past just above his head. By that time the bomb doors had been closed and our aeroplanes were able to fly out to sea, out-distancing the two German fighters which made a futile attempt to catch the Mosquitoes by chasing them for some 60 miles or so.

Only at one point in the whole operation were the Mosquitoes open to attack; that was when their bomb doors were open. Luck was with the Focke-Wulfs; otherwise they could never have taken off and intercepted the Mosquitoes, which succeeded, all the same, in wrecking the three small buildings which constituted the Nazi Party Headquarters in Oslo. The way in which these three buildings were identified among the many similar buildings in the town of Oslo was a notable feat in itself and exemplifies the best tactical use of the Mosquito against spot targets. Night bombers can rain an enormous weight of bombs on large areas but no night bomber can single out small targets, each of which has to be destroyed with the least amount of damage to surrounding, possibly civilian, property and with the smallest possible waste of high explosive.

The Mosquito's wood construction has proved as adequate to its task as any metal airframe. Fire does not appear to be causing any trouble, and although one or two fires have been known to occur as a result of short circuiting after radio sets have been damaged, they have not spread, and many of them have been put out by the Observers while still in flight. The wood stands up to cannon and machine-gun fire.

Manned by some of the most famous and competent air crews in the world, the de Havilland Mosquito light bomber squadrons of Bomber Command will continue to take advantage of every opportunity offered them to strike deep and hard into the resources of the Reich.—P.F.M.

MOSQUITO — *Fastest bomber in the world*

DE HAVILLAND

Clearing the skies for commerce

GREAT BRITAIN   CANADA   AUSTRALIA   NEW ZEALAND   INDIA   SOUTH AFRICA

The MOSQUITO

Fire power comprises four cannon and four machine guns in the nose.

Details of De Havilland Masterpiece : Day Bomber, Night Bomber, Day Fighter, Intruder, etc. : From Design Stage to Operational Service in 22 Months

IN trying to assess the merits of the design of the De Havilland Mosquito one should bear in mind that the general outlines of the machine were planned during the early weeks of the war. As soon as war was declared those responsible for the conduct of the De Havilland Aircraft Co., Ltd., realised that there would be a demand for a very fast bomber. The best liquid-cooled engine available at that time was the Rolls-Royce Merlin, and so the Mosquito was designed around two of these engines.

In view of the fact that the De Havilland Aircraft Co. had designed and built large numbers of wooden aircraft, it was perhaps natural that when it came to designing a military machine they should try to incorporate in their wartime type the forms of construction which had proved so successful in civil aircraft. As may be imagined, it was no easy task suddenly to switch ideas over from civil to military requirements, but the fact that the Mosquito went into operation 22 months after the design work had begun is proof that the designers of the company were able to do so very successfully. It is somewhat difficult to pick out from a large team the names of individuals who played a leading rôle in designing the new type, but in the first place credit must be given to Capt. Geoffrey de Havilland himself, one of our pioneer aircraft designers (he designed his first aircraft in 1908); Mr. C. C. Walker, chief engineer of the De Havilland Co. and a founder-director of the firm, has been associated with Capt. de Havilland ever since the days of the Aircraft Manufacturing Co. in the last war; Mr. R. R. E. Bishop, who is now chief designer; and Mr. R. M. Clarkson, assistant chief engineer and head of the Aerodynamics Dept. The actual detail work was, of course, spread over a very large technical staff, but these four

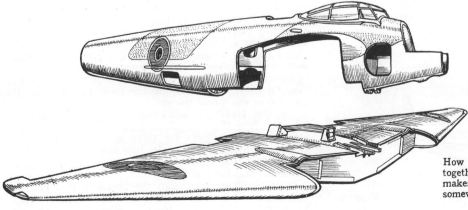

How fuselage and wing are brought together as units. The large cut-out makes the fuselage structure appear somewhat incomplete until it has been bolted to the wing.

men were in the main responsible for laying down the general lines upon which the Mosquito should be designed.

Performance and Production

The fundamental consideration in the design was that the machine should be as clean, aerodynamically, as possible. How well that ideal was attained is shown by the accompanying illustrations. Another basic consideration was that the aircraft should be capable of being produced not only by the De Havilland Co. but by a great number of other firms at home and overseas. The fact that the De Havilland Co. had had such long experience of wooden construction was, of course, taken into consideration, but in the *Flamingo* the firm had proved that it could turn out an extremely nice aircraft of all-metal construction, so that one may assume that other factors influenced the choice of wood as the structural material.

Among these one may mention the desire to get the design into construction as quickly as possible. When it comes to building a prototype it is a good deal quicker to make one in wood. The first idea was to design a bomber only, but it soon became obvious that the layout also lent itself to use as a twin-engined fighter, and thus the firm actually had to design two prototypes concurrently, a bomber and a fighter. In this connection it is not without interest to place on record the fact that the company was permitted by the Air Ministry and Ministry of Aircraft Production to proceed with this work *as a private venture*.

The choice of wood construction also meant that a great number of firms and a great deal of labour would become

"Bombs all gone." The bomb bay and its doors in the bomber version of the Mosquito The doors are operated by the two jacks which form an inverted vee.

available which had not previously been used to any great extent in the war effort. This was no small consideration, and has resulted in the manufacture of the Mosquito being more dispersed than probably that of any other aircraft type in the world. From the operational point of view it was thought that wood construction would have certain advantages such as buoyancy (Mosquitoes have floated for many hours) and ease of repair in case of damage. Almost any carpenter of average skill should be able to effect a satisfactory repair to many parts of the aircraft.

It is of interest to recall the fact that, in addition to the 400 or so sub-contractors making components for the Mosquito at home, the machine is now being built in quantities in Canada. The De Havilland Co. years ago started subsidiary companies throughout the Empire, in Canada, in Australia, in South Africa and in India. The latest De Havilland factory to get going overseas was that in New Zealand, which began work in 1939.

Basic Types

Reference has been made to what may be called the two basic Mosquito types: the bomber and the fighter. In addition to these, official reports from the various fronts have recently indicated that the Mosquito is in service as a day and night bomber, as a long-range day fighter, and as an intruder, all very highly successful. The type which we have chosen for our description and illustration this week is the fighter, the armament of which comprises four 20 mm. Hispano cannons and four 0.303 Browning machine guns All eight guns are mounted in the nose so as to give an extremely concentrated fire. The Mosquito bomber version carries 2,000 lb. of bombs with a fuel range which brings practically the whole of Germany within its reach. There is thus some justification for calling the Mosquito the most versatile first-line aircraft in the world to-day.

The De Havilland series number of the Mosquito is DH 98. It is the first military aircraft put into production by the firm since the DH 9 and DH 10 of 1918 or so. It is interesting to recall that the production of the DH 9 and DH 10 reached the impressive figure of 250 per month during the last war.

In its fundamental design the Mosquito is a mid-wing twin-engined monoplane with the wings tapering sharply to the tips and the two engine nacelles merging smoothly

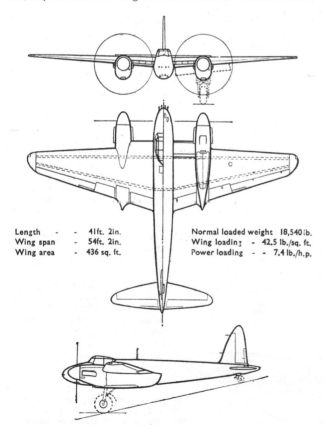

Length	-	-	41ft. 2in.
Wing span	-	-	54ft. 2in.
Wing area	-	-	436 sq. ft.

Normal loaded weight	18,540 lb.
Wing loading	-	42.5 lb./sq. ft.
Power loading	-	-	7.4 lb./h.p.

Three-view general arrangement drawings of the Mosquito fighter. Notable is the pronounced taper of the wings.

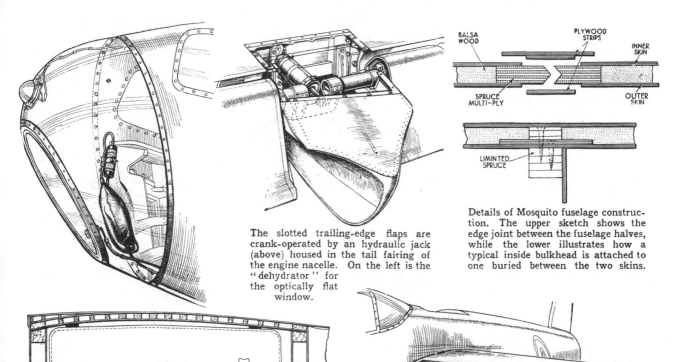

The slotted trailing-edge flaps are crank-operated by an hydraulic jack (above) housed in the tail fairing of the engine nacelle. On the left is the " dehydrator " for the optically flat window.

Details of Mosquito fuselage construction. The upper sketch shows the edge joint between the fuselage halves, while the lower illustrates how a typical inside bulkhead is attached to one buried between the two skins.

The top surface of the wing is spaced plywood skins connected by spanwise stringers. The tank doors, shown above, have balsa wood between skins, while the rest of the wing has single plywood covering on the under surface.

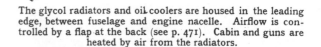

The glycol radiators and oil coolers are housed in the leading edge, between fuselage and engine nacelle. Airflow is controlled by a flap at the back (see p. 471). Cabin and guns are heated by air from the radiators.

into the surface of the wing. The nose of the fuselage is fairly short and almost in line with the airscrew spinners. Perhaps the feature which. more than any other, has helped to make the Mosquito so clean is the use of wing radiators housed in the leading edge between the engine nacelles and the fuselage. This placing of the radiators avoids the unsightly and drag-producing excrescences usually seen below the engine nacelles of a twin-engined aircraft fitted with liquid-cooled engines. The only excrescences on the engine nacelles are the forward-facing air intakes underneath and the flame traps on the exhaust. The wheels retract fully into the engine nacelles, and at the stern only a portion of the tail wheel projects when the wheel is retracted.

Smooth Airflow

The fuselage itself is equally clean. The only excrescence here is the small roof over the cockpit. Had the machine been designed with gun turrets the beautiful clean lines would, of course, have been entirely spoilt, and the machine was designed for speed rather than for defensive gun power. Presumably rearward firing guns *could* be mounted, but

they would certainly spoil the appearance of the machine and would undoubtedly reduce the speed very considerably.

In the structural design of the Mosquito, full use has been made of the experience obtained by the De Havilland Co. in the design and construction of the Comet which won the England-Australia race and of the Albatross air-liner, although the latter was a four-engined type. New methods of applying wooden construction to modern aerodynamic forms were evolved in the case of these two aircraft, and much of that experience has been applied in the construction of the Mosquito.

The fuselage of the machine is a wooden shell, the inner and outer skins of which are plywood, and sandwiched in

A Mosquito out looking for trouble. In spite of the engine nacelles the crew obtain a good view in essential directions.

between which is a layer of balsa wood. The purpose of the balsa wood is, of course, to stabilise the relatively thin plywood skin.

Split Construction

In conformity with modern practice, the fuselage is built on the principle of split construction. This was advisable not merely because it enabled the installation of equipment to be carried out so much more readily, but was rather essential during the course of manufacture because otherwise the locking of internal bulkheads and formers to the external members, which are worked into the space between inner and outer skins, would be very difficult indeed, not to say impossible. As it is, each bulkhead and former, which is made up of laminated strips, has it counterpart on the outside, where thin laminated strips are screwed through the inner skin to the inner bulkhead. Where local strength demands such a course, structural members are worked into the space between the two skins so that reinforcement is provided in this manner. Such members lie, of course, in openings left in the balsa-wood packing.

Incidentally it should be mentioned that the fuselage

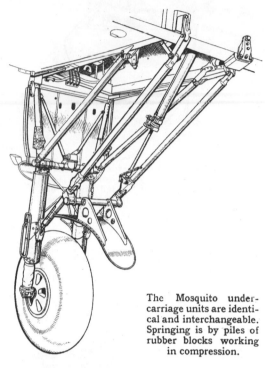

The Mosquito under-carriage units are identical and interchangeable. Springing is by piles of rubber blocks working in compression.

is split along the vertical centre plane. When the equipment has been installed, or at least as much of it as it is possible to install before the fuselage is assembled, the two halves are brought together. The method of securing the free edges of the two halves is rather interesting. The edge of one shell-half is in the form of a vee-section groove. The corresponding edge of the other shell-half is roof shaped. When the two are brought together they thus form a good shear joint. It should be mentioned that the extreme free edge of the fuselage shell-half is a plywood strip. The joint is finally completed by internal and external plywood strips covering the joint.

The rear fuselage portion was a very easy problem because there are no cut-outs in this, or at least only one, i.e., the access door near the bottom on the starboard side. In the front part of the fuselage, however, very different problems arose. Here the wing had to be accommodated, and the way in which this is done is rather interesting.

Owing to the fact that the wing is about mid-way up the fuselage, and that in front of it at the top there is a large cut-out for the cockpit, there is very little wooden material left wherewith to tie the fuselage shell to the wing spars. What may be termed a short longeron placed slightly below the mid-way centre line of the fuselage carries at its forward end a metal fitting which picks on a corresponding fitting on the bottom flange of the rear spar.

The concentrated load from the front spar to the fuselage shell is taken care of by a sloping and twisted member which carries at its forward end a metal fitting picking up on a corresponding fitting on the top flange of the front spar. This sloping longeron, as it might be termed, distributes the concentrated load from the fitting to the shell and to one of the bulkheads of the fuselage. The attachment is finally completed by numerous bolts connecting the fuselage shell to specially strong wing ribs.

The side panels and ventral doors below the wing are both fitted after the wing has been assembled to the fuselage structure.

The One-piece Wing

The wing is a one-piece wooden cantilever with built-up spars having laminated spruce flanges and plywood birch webs. The ribs between the spars have spruce booms and plywood webs.

In the wing covering the Mosquito differs somewhat from the Comet and Albatross. The lower covering is plain birch plywood, except for the doors over the tank bay. The top skin, however, is different in that there are two skins, spaced some distance apart, and instead of the balsa wood packing between them, as used in the fuselage, they have stringers running spanwise. These stringers are fairly closely spaced and serve to stabilise the two skins. Otherwise the outer skin, being further from the neutral axis, would take a greater load than the inner. Generally speaking, the leading edge is built as a separate unit and screwed on when the various leads, controls, etc., have been mounted on the front face of the front spar.

The tank doors, on the other hand, which have to transmit

Mounting of one of the Rolls-Royce Merlin XXI engines on the wing of the Mosquito. The engine bearers are carried by the fixed portion of the undercarriage structure.

stresses, are of the same fundamental construction as that of the fuselage, that is to say, with two skins of plywood with a packing of balsa wood between them. The edges of these tank doors are secured to angle-section strips bolted to the ribs and spars by numerous Simmonds stop nuts.

The trailing-edge portion of the wing is of generally similar construction, but the ailerons are of light alloy construction with sheet covering of the same material. The trailing-edge flaps are of the slotted type and are of wooden construction, with spruce ribs and plywood covering. The dihedral angle of the wings begins at the fuselage side, so that there is only a short length of spar which is horizontal. All the rest of the wing is set at dihedral angle. The wing tips are made up as separate units screwed on to the main wing when the navigation lights have been installed.

The two undercarriage units are identical and interchangeable. They comprise the usual structure of steel tubes, with the front legs hinged at their tops and the rear legs having a " breakable " joint operated by an hydraulic jack. In the compression legs themselves, however, an unusual feature is found. In place of the usual hydraulic or oleo-pneumatic undercarriage leg, De Havillands have made use in the Mosquito of a system which was found to work very satisfactorily in the Moth, but which had not previously been attempted on a machine of the weight of the Mosquito. The shock-absorbing medium is a pile of rubber blocks working in compression. It will be realised that this avoids the

Split construction is employed in the Mosquito fuselage to facilitate installation of equipment.

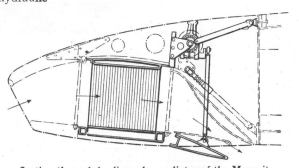

Section through leading-edge radiator of the Mosquito. The airflow is controlled by the flap at the back, the front being open. A perspective view is given on p. 469.

need for very accurately machined cylinders and thus effects a very great saving in time during manufacture. These struts have been found to be perfectly satisfactory in service. The tyres and wheels are Dunlops, fitted with pneumatic wheel brakes. The retractable tail-wheel unit has a Dunlop wheel with Acta tyre which conducts electricity into the ground.

The two Merlin XXI engines are mounted on engine bearer frames in the form of steel tubes bolted to the front spar of the wings and to the fixed portion of the undercarriage structure. Starting is by an electric motor and booster coil, and priming is by a Ki-gas pump in each engine nacelle. The engine cowls merge neatly into the wing surface, and it will be observed that in conformity with modern practice the nacelles are mainly below the wings. The Merlin engines drive De Havilland three-bladed hydromatic fully feathering airscrews.

The Fuel System

For normal ranges all the fuel is carried in wing tanks, of which there are eight, four on each side. The two outer tanks each have a capacity of 24 gallons, the next two a capacity of 34 gallons each, while the capacities of the inner tanks are $65\frac{1}{2}$ gallons and 78 gallons respectively. The total petrol capacity is thus 403 gallons. For particularly long-range work extra tanks may be fitted in the fuselage. These have a capacity of 150 gallons, bringing the total for long-range work up to 553 gallons.

Actual performance figures may not be quoted, but in view of the fact that the wing loading is over 42 lb./sq. ft. one may assume that the Mosquito is capable of a speed not very

A Mosquito fighter flying over the Mediterranean in the neighbourhood of Malta.

far short of 400 m.p.h., and on the assumption that the cruising speed is somewhere round about one-half of that figure, the total tankage capacity of the machine should give a range of at least 1,200 miles, probably a good deal more.

One result of the remarkably clean design of the Mosquito is that even with full load the aircraft will not only maintain height but will climb on one engine without running that engine full out.

Single-engine Performance

In spite of the fact that there is but a single fin and rudder, pilots find no difficulty in keeping the machine straight with one engine stopped and its airscrew feathered. Not only so, but turns are made quite readily "against" the working engine. Young Geoffrey de Havilland has an intriguing demonstration in his repertoire. With one engine stopped he will do an upward roll from nearly ground level, showing that the Mosquito has ample power.

While on the subject of power it may be mentioned that the Merlin XXI engine is identical with the Merlin XX described very fully in our issue of February 26th, 1942, except for the reversed cooling flow connected with the leading-edge radiators.

Reference has already been made to the armament of the fighter version of the Mosquito, which includes four 0.303 Browning machine guns in the nose and four Hispano 20 mm. cannon mounted underneath the forward part of the fuselage. These gun positions have the advantage that not only are they so placed as to give a remarkable concentration of fire, but when the aircraft is on the ground they are very accessible for overhaul and servicing. The guns are fired electro-pneumatically, there being an air compressor in the port engine nacelle. This compressor, in addition to serving the guns, also operates the pneumatic wheel brakes. There are two vacuum pumps, one in each engine, which together operate the instrument flying panel. The pneumatic system of these pumps is so arranged that in the event of one breaking down it is automatically isolated and the other pump is able to carry on.

The cockpit for pilot and observer is entered through a door on the right-hand side below the wing. A telescopic ladder is carried which gives access to the cabin. Hot air from the cooling system of the engines maintains the temperature in the cabin at a convenient degree. A control behind the pilot's seat enables the temperature to be regulated to suit the weather conditions in which the aircraft is flying.

The Mosquito tail wheel is retractable. The extreme stern portion of the fuselage is formed by a conical fairing.

The pilot occupies the seat on the left, while the observer's seat is on the right of the pilot and very slightly behind him. In the bomber version there is a prone bomber's position below the seats and in front of them.

Reference has been made to the different versions of the Mosquito. One of these claims the proud distinction of being the fastest aircraft in operation in the world.

In a long life the De Havilland Aircraft Co., Ltd., has produced many fine types of aircraft, but the Mosquito can truly be said to be the firm's masterpiece.

A sight which the enemy fears, but which friends in occupied countries welcome. A formation of Mosquitoes.

MOSQUITO

Fastest aircraft in service in the world

Most versatile first-line aircraft

From first conception to operational
service in 22 months

World-wide dispersal of manufacture

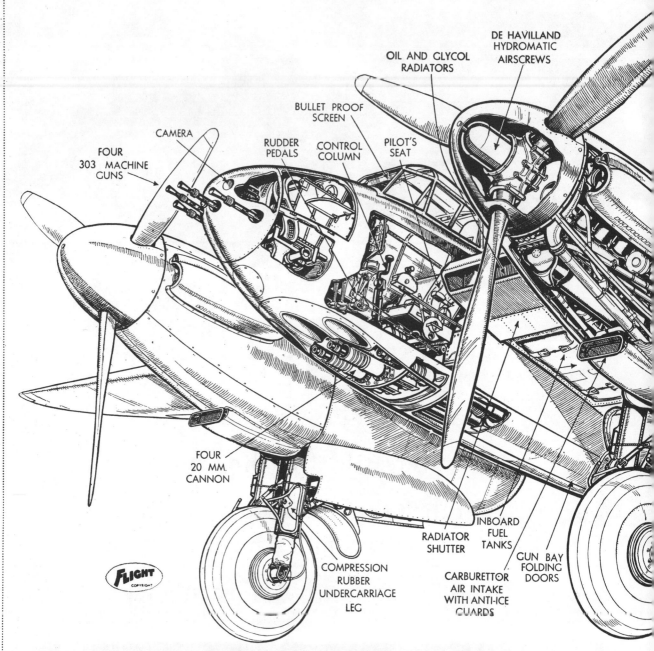

FOUR
303 MACHINE
GUNS

CAMERA

RUDDER
PEDALS

CONTROL
COLUMN

BULLET PROOF
SCREEN

PILOT'S
SEAT

OIL AND GLYCOL
RADIATORS

DE HAVILLAND
HYDROMATIC
AIRSCREWS

FOUR
20 MM.
CANNON

COMPRESSION
RUBBER
UNDERCARRIAGE
LEG

RADIATOR
SHUTTER

INBOARD
FUEL
TANKS

CARBURETTOR
AIR INTAKE
WITH ANTI-ICE
GUARDS

GUN BAY
FOLDING
DOORS

DE HAVILLAND MOSQUITO F.II (FIGHTER)

In this copyright drawing by our chief artist, Mr. M. A. Millar, the general layout of the Mosquito is clearly revealed. The primary structure is of wood, largely in the form of plywood. Other constructional details will be found on the preceding pages. Notable features are the leading-edge radiators and the use of compression rubber blocks in the undercarriage legs. The crew of two comprises pilot and navigator, who are seated side by side. The machine guns and cannon are fired electro-pneumatically by switches on the control column. The guns and cabin are heated by air from the radiators.

DATA

Two Merlin XXI Engines

Duty : Two-seater long-range fighter
Crew (2) : Pilot and observer

Length o.a.	-	-	-	-	41ft. 2in.
Wing span	-	-	-	-	54ft. 2in.
Wing area (gross)	-	-	-	436 sq. ft.
Root chord	-	-	-	-	12ft. 3in.
Tip chord	-	-	-	-	3ft. 10in.

Aspect ratio	-	-	-	7
Max. fuselage depth	-	-	5ft. 5.5in.
Max. fuselage width	-	-	4ft. 5in.
Wheel track	-	-	-	16ft. 4in.
Normal loaded weight	-	18,540 lb.
Wing loading	-	-	42.5 lb./sq. ft.
Power loading	-	-	7.4 lb./h.p.

The Rolls-Royce Merlin XXI Engine

Bore	-	-	-	-	5.4in.
Stroke	-	-	-	-	6.0in.

Capacity	-	1,649 cu. in. (27 litres)
Max. power	-	-	1,250 b.h.p.

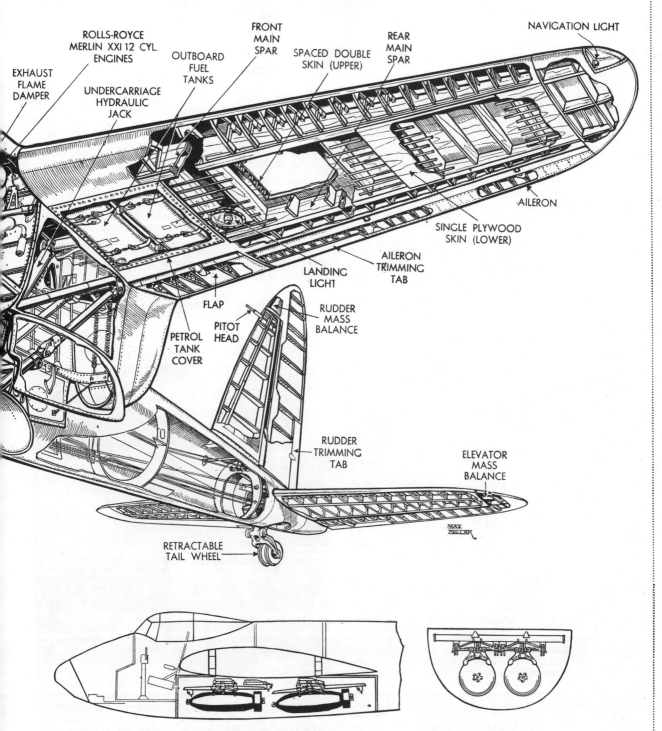

EXHAUST FLAME DAMPER

ROLLS-ROYCE MERLIN XXI 12 CYL. ENGINES

UNDERCARRIAGE HYDRAULIC JACK

OUTBOARD FUEL TANKS

FRONT MAIN SPAR

SPACED DOUBLE SKIN (UPPER)

REAR MAIN SPAR

NAVIGATION LIGHT

AILERON

SINGLE PLYWOOD SKIN (LOWER)

AILERON TRIMMING TAB

LANDING LIGHT

PETROL TANK COVER

FLAP

PITOT HEAD

RUDDER MASS BALANCE

RUDDER TRIMMING TAB

ELEVATOR MASS BALANCE

RETRACTABLE TAIL WHEEL

Side and front elevations of the bomb bay of the bomber version of the Mosquito. The bomb load is 2,000lb. with sufficient fuel to reach most parts of Germany. Four 500lb. bombs are shown. Apart from the arrangement of armament, the fighter and bomber versions are very similar.

[" Aeroplane " photograph

BORN TO EXCEL.—A de Havilland Mosquito fighter climbs to clear the aerodrome control tower after diving to within a few feet of it. Such flying is possible only when the pilot has complete trust in his aeroplane. In the Mosquito, high speed is accompanied by fine handling qualities and by a power of manœuvre which is rare in two-motor aircraft.

A FIGHTING MOSQUITO SQUADRON

WITH MOSQUITO fighters coming off the assembly lines in swarms, the Air Ministry has lately had the pleasant task of selecting squadrons to fly them. One squadron upon which the choice fell formerly flew Defiants. It is, in fact, the squadron which shocked the Luftwaffe over Dunkirk in May, 1940, by shooting down a whole Staffel of Messerschmitt Me 110s before breakfast and ending the day with the record bag of 37 confirmed victories.

The squadron received its Mosquitoes nearly a year ago and has now attacked and damaged 43 locomotives, shot down two Junkers Ju 88s and damaged one and probably damaged another; probably destroyed a Heinkel He 111, damaged a Do 217, shot up 10 power stations and electrical transformers, wrecked a number of lorries, made some 60 individual patrols over the Bay of Biscay and some 70 day and night intrusions, mostly into France.

Pilots had no trouble in acquiring the Mosquito technique. Those who were not accustomed to twins were given brief spells on Oxfords; those who were, studied the Mosquito manual and forthwith took off. This was not the hazardous adventure that might be imagined. There are no tricks to learn, no vices to master, no disconcerting habits to be watched. In the Mosquito, the pilots discovered the responsiveness of the Tiger Moth, and in it powerful motors, swift acceleration and great speed, not added dangers, but new and exhilarating sources of pleasure.

They soon found themselves beneath the Mosquito's spell. To them, it is well-nigh flawless. They admit of no equivalent and would refuse any substitute. They eagerly accept every mission that comes to them because they know that they have every chance of fulfilling it and returning. They rate it an honour that they were among the first to receive the Mosquito and they are prepared, on occasion, to show their skill to their friends, without " effects," as readily as they are to the enemy with them. They welcomed the opportunity to " show us the works " when we visited them recently.

In diving on the control tower and shooting it up with silent guns they showed faultless judgment. At their speed, an error of a hair's breadth or of the smallest conceivable portion of the time would have been fatal, but with all the ease, grace and carelessness imaginable, they deftly lifted themselves up and over the heads of the roof-top watchers and swung away on steeply banked wings into the cold grey sky of a Spring afternoon. It was not a crazy exhibition, but flying of a kind that makes admiration breathless.

[" Aeroplane " photograph

READY FOR MORE EXPLOITS.—Pilot and navigator with a Mosquito fighter which has already taken toll of seven locomotives by day and one by night, and shot up a power station. This pilot (perched on the centre section) had the further distinction of bringing a Mosquito 400 miles home on one engine.

They had used most of the runway for the take-off, preferring, they said, the asset of speed to that of height. A little flap would have made them airborne earlier, but they refrained from using it. Nor were the motors fully extended; there still remained a good margin in hand. Full throttle running is uneconomic and shortens a motor's life. Over enemy territory a few minutes later, the Mosquito might want all that the motor can give, and worn parts mean lost power.

The pilots always fly with a navigator. A sortie may send a Mosquito 600 miles from its base—perhaps at tree-top or wave-crest height all the way. Piloting, then, is a full-time occupation. The man at the helm must have his eyes outside the cockpit; a momentary glance at a map or a compass might mean disaster. The navigators are chosen for the speed and accuracy of their calculations.

At first some of the navigators found the Mosquito's speed excessive. One, bound at night for a well-defended aerodrome in France, announced to his pilot that they would soon be there. The aerodrome they were seeking at once corrected the navigator by filling the sky around with a great fury of "flak." It was the defence, not the attackers, who scored the surprise.

Speed is but one of the Mosquito's virtues. Armament is another. In the belly are four cannon; in the nose, four machine-guns. The machine-guns are normally reserved for "soft skin" targets such as motor lorries, the wooden huts of military camps, and enemy troops. The cannon disable locomotives, prise open armoured cars, smash canal lock gates, wreck dynamos and transformers, and destroy aircraft. The choice of cannon or machine-guns, or both, is made through a selector switch in the cockpit.

In a dog fight the fighter version of the Junkers Ju 88 is poor sport for the Mosquito. Three Mosquitoes caught a pair of them over the Bay of Biscay searching for Coastal Command bombers on anti-submarine patrol. In the brief engagement which followed the two enemy fighters were shot down with little ceremony. All three Mosquitoes had a hand in the slaughter, but a pilot who was there said that one Mosquito could have coped with the situation. That remark was not a boast. Nor was it line-shooting. It was a modest tribute to a great aeroplane expressed as a fact.

Any attempt here to describe all the virtues of the Mosquito would be merely to send a faint echo bounding off the mountain of praise already heaped upon it by those who fly and fight with it. To quote the epic 400 miles' flight home on one motor, made by the squadron's commanding officer; to describe the lightness of the damage caused by ground fire to the wood structure, and to record the crews' satisfaction with their cockpit would be merely to extol qualities already extolled.

Of all modern aeroplanes, the Mosquito must be confessed the fittest subject for commendation. In its shapely fuselage it compounds the elements of fury, havoc and destruction. In its speed it stretches the enemy's defences to the utmost. In its range it lengthens Fighter Command's arm, and in its construction it makes the smallest calls upon our precious stocks of metal.

To talk with the pilots and navigators of this squadron is to learn the meaning of high morale. Suggest that their work is fraught with danger and they smile. Hint that they need frequent rests from operations to recover their nerve and they are scandalised. They are eager to be off on sorties and would have taken a dim view of the Press visit which kept them on the ground for the better part of the day had the weather been right for locomotive hunting.

Their ground crews, too, are in high humour. They await the return of their Mosquitoes, paint pot in hand it seems, ready to emblazon the nose with the symbols of the latest targets. No academician ever worked with greater zest than the man who paints the little white and brown railway engines —white for daylight victims, brown for night—for all the squadron to see. When the nose of the machine carries the picture of a locomotive yard—and several do—the ground crew is apt to swagger.

If the same spirit pervades other Mosquito fighter squadrons —and it unquestionably does—then, indeed, the enemy has plenty to fear. These young fellows, keen and aggressive and working in paired partnerships, are proud of their squadron and proud of the work it is doing. And in the Mosquito fighter they have a mighty foundation for their pride.—s.v.

["Aeroplane" photograph

ELEMENTS OF DESTRUCTION.—The Mosquito fighter packs four 20 mm. cannon in the fuselage and four .303 in. machine-guns in the nose.

["Aeroplane" photograph

A FAMOUS SQUADRON.—The Defiant squadron which won undying fame at Dunkirk in 1940 has now been re-equipped with D.H. Mosquito fighters. Here pilots and navigators are grouped in front of a Mosquito. One of them is an original member of the squadron and has been away from it only 10 months since the War began.

De Havilland Mosquito

(Fighter Version)

POWER PLANT, TWO ROLLS-ROYCE MERLIN XXI ENGINES,
EACH DEVELOPING 1,260 B.H.P. AT
12,250 FT. ALTITUDE.

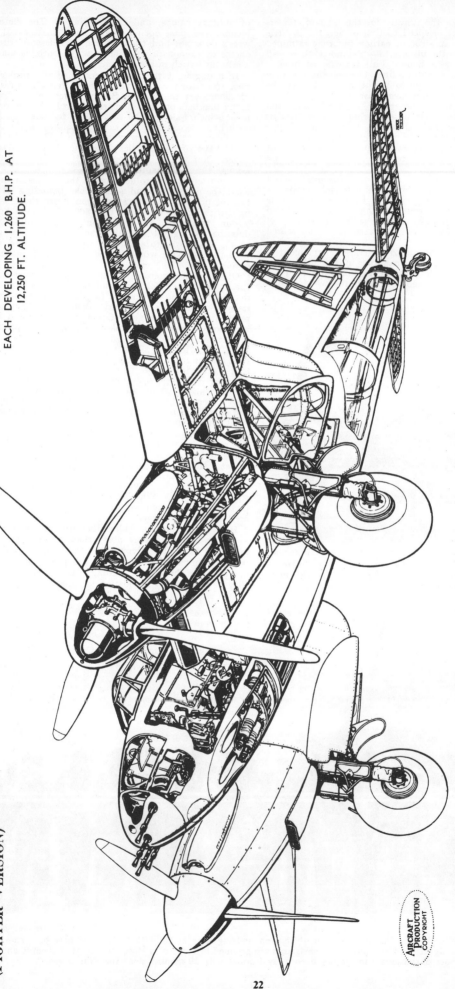

The de Havilland Mosquito is unique as being the first modern operational aircraft of all-wood construction to go into service anywhere in the world. It is also the only British aircraft now in regular service to have been designed, produced in quantity and put into service since the beginning of the war. From the commencement of design until the machine went into operational use a period of only 22 months elapsed.

There are two basic versions of the Mosquito, bomber and fighter, with several variants of the two types, including day and night bomber, long-range day fighter and intruder. The two basic versions differ mainly in the arrangement of the nose and the bomb-bay. The fighter version shown here has an armament of four .303 Browning machine guns in the nose and four 20 mm. cannon arranged under the fuselage, while the bomber carries a

maximum bomb load of 2,000 lb. in the form of two pairs of 500-lb. bombs.

The most outstanding structural characteristic is the fuselage, built as a balsa-wood-ply-wood sandwich, with an integral stiffening structure giving a remarkably clean interior. Wing construction is more conventional and is based on two box-spars of the usual type. The skin covering is unusual, consisting of plywood, reinforced with closely-spaced square-section stringers. Over the top surface the stringers are sandwiched between a double skin. These and other details can be seen in the sectional drawing above. A particularly interesting feature is the use of rubber for the compression strut of the undercarriage. Production processes on the bomber version of this outstanding aircraft are described in the accompanying article.

DE HAVILLAND MOSQUITO

The World's Fastest Aircraft in Production : Moulded, Split-fuselage Construction : Wing-spar Manufacture

By WILFRED E. GOFF

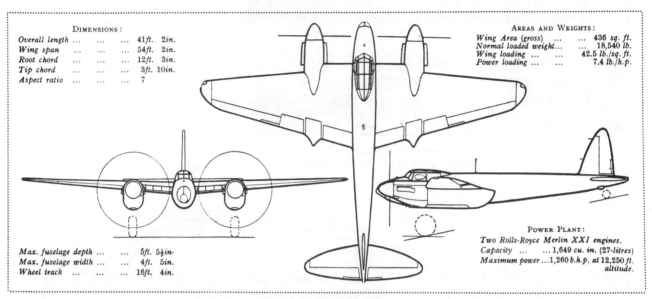

<table>
<tr><td colspan="2">DIMENSIONS :</td></tr>
<tr><td>Overall length ...</td><td>41ft. 2in.</td></tr>
<tr><td>Wing span ...</td><td>54ft. 2in.</td></tr>
<tr><td>Root chord ...</td><td>12ft. 3in.</td></tr>
<tr><td>Tip chord ...</td><td>3ft. 10in.</td></tr>
<tr><td>Aspect ratio ...</td><td>7</td></tr>
</table>

AREAS AND WEIGHTS :	
Wing Area (gross) ...	436 sq. ft.
Normal loaded weight...	18,540 lb.
Wing loading ...	42.5 lb./sq. ft.
Power loading ...	7.4 lb./h.p.

Max. fuselage depth ...	5ft. 5½in.
Max. fuselage width ...	4ft. 5in.
Wheel track ...	16ft. 4in.

POWER PLANT :
Two Rolls-Royce Merlin XXI engines.
Capacity ... ...1,649 cu. in. (27-litres)
Maximum power ...1,260 b.h.p. at 12,250 ft. altitude.

Fig. I. General arrangement and data for the de Havilland Mosquito.

DURING the last two decades wood as a structural material for first-line aircraft has been passing through a period of almost total eclipse. Even during the last five years or so, however, when a good deal has been heard, principally from the United States, of so-called plastic aircraft and other modern applications of wood construction, little or nothing has been published concerning British activity in this sphere. It has, nevertheless, been left to this country to give the most convincing demonstration of its possibilities by producing in the Mosquito the first modern all-wood first-line aircraft to go into service.

It is particularly appropriate (though probably inevitable) that the de Havilland Aircraft Company should have been responsible for this achievement. This firm has been very faithful to wood, though in metal aircraft design it has shown its quality also by producing the Flamingo, one of Britain's very few competitive types of modern civil transports, which appeared a year or so before the war.

Apart from the Company's long experience in wood construction, however, there were several considerations which led to its adoption. With wood it was possible to cut down the initial design stages and to build the prototype machine much more quickly than would have been possible if metal construction had been used.

As a result, it was possible to put the machine into production much more rapidly, a matter of prime importance in the circumstances. Also, the use of wood avoided imposing additional strain upon metal supplies and, further, made it possible to employ a class of skilled labour made available by the restricted activities of the woodworking trades generally.

Many coachbuilding and furniture factories have been turned over to sub-contract work on the Mosquito. In fact, apart from the Company's own dispersal system of factories, which is very extensive, there are some 400 sub-

AUTHORITATIVELY described as the fastest operational aircraft in the world, the de Havilland Mosquito is the first modern first-line machine of all-wood construction to go into service. It is a twin-engined mid-wing cantilever monoplane and is built in two basic versions —bomber and fighter. Of these two there are several variants, including day and night bomber and intruder types. The bomber version will carry a bomb load of 2,000 lb.

Structurally, the most outstanding feature of the aircraft is the fuselage, built on the balsa-plywood sandwich principle introduced by the de Havilland Company prior to the war and first used in the four-engine Albatross civil transport. The power plant consists of two Rolls-Royce Merlin XXI engines, each developing 1,260 b.h.p.

contractors engaged upon the production of the machine. It is also in large-scale production in Canada. Another advantage of the wood construction is that serviceable repairs can readily be effected by carpenters of average skill.

The Mosquito is a direct product of the war and it is a tribute to the vision and foresight shown in the original conception, and to the method of construction adopted, that the aircraft was put into service less than two years after the commencement of design in the early weeks of the war.

This feat is more creditable in view of the fact that the de Havilland Company had not a long-standing tradition in military aircraft design. To have produced the fastest operational machine in the world, as it were from scratch, is no mean achievement in this country, where excellence in this field of design is the rule rather than the exception.

Apart from its superlative performance, the Mosquito embodies some remarkably interesting features of construction, chief of which is the plywood and balsa-wood sandwich of the fuselage. This type of construction was first used in the de Havilland four-engine Albatross civil transport and must rank as one of the simplest and quickest methods yet devised of producing a monocoque type of unit. The division of the fuselage into halves is made along its vertical centre line and considerably simplifies subse-

Fig. 2. In the first stage of fuselage assembly the bulkheads and other members of the internal structure are located in slots in the mould.

quent assembly operations, particularly in those stages where services and equipment are installed.

Building the Fuselage

Each half of the fuselage is built n the horizontal position with the joint line downward. A male form or mould is used, made of mahogany or concrete and shaped to the interior form of the fuselage. The procedure followed is reminiscent of that used for the production of moulded plywood units, although neither heat nor pressure-chamber is used to obtain the required curvature.

The internal stiffening structure is produced integrally with the covering or skin. The first operation is the location of these structural members in slots and recesses in the mould (Fig. 2). Six of the seven bulkheads are positioned in transverse slots, which locates them longitudinally, while laterally they are retained by

stops against which they are pressed inward by the application of the skin. The bomb-aimer's floor-bearers in the nose are also located in slots in the mould.

Wing Attachment Fittings

Bomb doors and fuselage lower side-panels are made integrally with the half-fuselage and are subsequently cut out to produce separate units. Recesses for the internal stiffening members of these components are also provided on the mould. A point of particular interest is that the main attachments for the wing are also moulded into the half-shell assembly (Fig. 3). These fittings are attached to their wooden support member, and are located on the mould by pins inserted through the fuselage pick-up lug and by a slot which receives the wooden member itself.

The inner plywood skin of three-ply birch, 1.5 and 2 mm. thick, is next applied over the structural members.

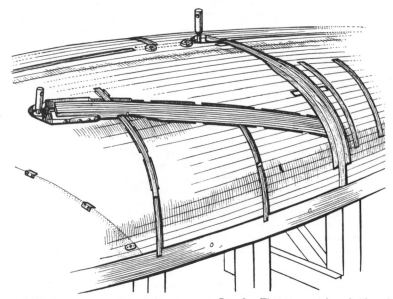

Fig. 3. The two starboard-side wing pick-up fittings located with their support members in the fuselage mould.

Fig. 4. The inner fuselage skin and the between-skin structural members being fitted in the second stage of fuselage assembly.

Some of the skin sections, such as that for the bomb door, are preformed, but, in general, this is not found to be essential, and the skin over the rear portion is applied in relatively large panels, which are scarfed from smaller sections.

The vertical joints between panels are arranged to fall between and not less than 6in. away from the bulkheads. Longitudinal joints, on the other hand, are made on spruce stringers of the between-skin stiffening structure. Narrow strips of skin, scarfed and glued together, are used to cover the nose where sharper double curvatures are encountered.

Broad, flexible steel bands cramp the skin down over the mould. The cramps are tightened by turnbuckles which work on the union-nut principle and engage opposite-hand threads on studs attached respectively to the end of the steel band and to the jig base. Rotation of the turnbuckle by a tommy-bar draws the steel bands down very hard over the skin surface and exerts a very high pressure for the bonding operation. The cramps are spaced very closely, in fact, their edges almost touch (Fig. 6), so that the pressure is applied over the whole surface of the skin. Through perforations in the steel bands the skin is stabbed to permit excess cement to ooze or "bleed" through from the inner surface during the setting period.

Between-skin Structure

Between the inner and outer skins there is a stiffening structure of laminated spruce. Laminated spruce strips are used at the bulkhead stations

Fig. 5 (above). Fitting the balsa-wood filling which forms the centre portion of the sandwich.

Fig. 6 (left). Flexible steel-band cramps are used between the various stages of assembly to give the required pressure for bonding.

in the form of battens or strips of varying width (Fig. 5) and the entire area between the between-skin members is filled, the thickness being about $\frac{3}{8}$in. After the whole of the balsa has been dry-fitted, it is removed, cement applied to the under surface and to the inner skin, and the steel-band clamps are again placed in position and tightened down. The steel bands are lightly greased to prevent the adhesive from sticking to them and so pulling the balsa when they are removed. When they are taken off after the setting of the cement, the balsa is surface smoothed to the contours of the fuselage. As in the case of the first skin the balsa is stabbed for bleeding of the cement. Fitting and cementing of the outer skin panels,

and are screwed through the inner skin to the bulkheads inside the fuselage. Longitudinal spruce members provide additional strength and reinforcement where required. Spruce doubling members or coamings are also applied to enclose the areas where doors and other apertures are subsequently cut out. These members are located by

pins inserted through them into the fixture and are first dry-fitted and then cemented in position.

Balsa-wood Filling

In the next stage the balsa-wood centre filling is fitted, which serves the purpose of stabilising the very thin inner and outer skins. It is applied

Fig. 7. Drilling the interior of the fuselage for the ferrules through a wooden template jig.

Fig. 8. The ferrules in position with some of the fittings mounted.

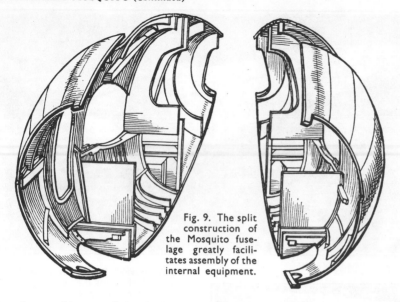

Fig. 9. The split construction of the Mosquito fuselage greatly facilitates assembly of the internal equipment.

also of 1·5 and 2 mm. three-ply birch, is performed in the same way. Finally, the half-shells are cleaned out before being sent to the assembly department.

Fuselage Assembly

During the first stages of assembly the shells are held in a simple vertical cradle on casters, turnbuttons being used to retain them in place. A rather attractive feature of the Mosquito construction is the use of " ferrules " for the attachment of fittings to the interior surface of the fuselage skin. These ferrules consist of a plywood disc of liberal diameter mounted on an ebonite or wood plug in which a threaded metal (brass) ferrule is set so that its open end appears in the centre of the disc. To mount the ferrules, holes are drilled in the fuselage skin from the inside, of a sufficient depth to receive the plugs, but not so deep as to break through the outer surface.

The first operation on the half-shell in the assembly shop is to drill these holes. Large wooden templates (Fig. 7) are used for this purpose, cramped or tacked to the shell and located from its edges or from the internal structural members of the fuselage. A relatively high degree of accuracy is necessary in this operation, as many of the fittings and items of equipment have jig-drilled brackets or feet for attachment to the fuselage. With the template in position, the holes are spotted through with a centre-bit to start the finishing tool,

which is in the nature of a hollow-mill or trepanning cutter and is used to finish the holes to depth when the template has been removed. The plugs are then glued into the holes and the ferrules finally attached by tacks through the plywood disc.

In the next stage of assembly, the bomb-door and side-panel are cut out of the fuselage side and sent to the wood detail shop, where they are completed and jig-fitted with all-metal parts such as bomb-door hinges.

Installing Equipment

In the half-shell stages of assembly as much of the interior equipment is installed as possible. In the earlier stages the remaining portions of the external structure, including the cockpit and bomb-aimer's floors, are assembled, and some of the armour plating on No. 2 bulkhead is fitted before the spray-painting of the interior. Surfaces where a cemented wood-to-wood joint is required are masked with adhesive tape before the paint is applied. Electrical bonding of the metal fittings in the fuselage with copper strip follows spray-painting.

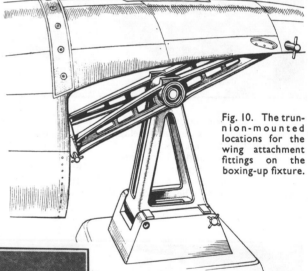

Fig. 10. The trunnion-mounted locations for the wing attachment fittings on the boxing-up fixture.

Later stages of half-shell assembly are concerned with the progressive assembly and installation of equipment and services. The rear floor is fitted, rudder and elevator-operating linkage is mounted in the cockpit floor and a commencement is made with the electrical wiring, and with the plumbing of the oxygen and hydraulic systems. A point of interest is the simplification of assembly by arranging the control cable runs down the port side of the fuselage and the hydraulic plumbing as far as possible

Fig. 11 (left). Equipment assembly proceeding on two half-fuselage sections prior to boxing-up. The trunnion mounting for the wing attachment pick-ups is seen in the centre.

Fig. 12. View, from the tail end, of the fuselage on the boxing-up fixture. Note the wood cramps on the after portion.

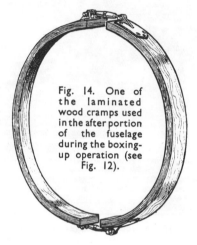

Fig. 14. One of the laminated wood cramps used in the after portion of the fuselage during the boxing-up operation (see Fig. 12).

the side panel below it, a temporary jury strut is inserted across the gap to maintain the rigidity of the structure and prevent damage or distortion during assembly. Aft of the wing gap, the halves of the fuselage are cramped together by wooden bands, shown in Figs. 12 and 14, which are drawn together by a screw-turnbuckle fastening similar to those used for the fuselage-moulding cramps. Forward of the gap, a cramp is placed across the rear of the nose portion on the level of the pilot's floor to draw the halves together to the required overall dimension. Plywood gussets are used to make the joints between the half-sections on the centre line of each bulkhead.

The actual form of the joint between the skins of the two half-shells is of interest. The inter-skin member at the free edge of the starboard shell is shaped to a V-section which fits into a V-shaped recess in the corresponding member of the port shell. Both inside and outside, the skin is rebated or recessed on each side of the centre

down the starboard side. The control column is also mounted in the port half-shell and connected to the rudder and elevator linkage before the joining stage is reached.

Joining the Fuselage

About 60 per cent. of the installation work is completed in the half-shell stages under conditions of maximum accessibility. This represents a very large saving in assembly time by comparison with that which would be required if the work had to be done in

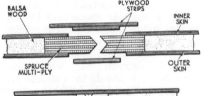

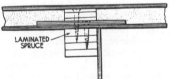

Fig. 13 (above). Details of the joint between the fuselage halves and (below) a typical section through the balsa plywood sandwich at one of the bulkhead stations.

as datum faces for setting the fuselage level both longitudinally and transversely by clinometer.

At the aft end, the fuselage is supported in a cradle, divided on the centre-line into halves, which can be drawn apart to simplify the mounting of the fuselage for joining the two shells. The nose section is supported from beneath on cradles mounted on screw-jacks which permit of vertical adjustments for levelling by clinometer readings taken from datum points inside the nose. A check is maintained on the position of the centre-line of the fuselage by plumb-bobs suspended from the mid-points on the bulkheads and aligned with datum markings let into the floor.

At this stage, because of the wing gap in the fuselage and the absence of

a closed fuselage in which only a very limited number of operatives could work at any given time. The operation of joining the halves of the fuselage, known as "boxing-up," is done on the fixture shown in Figs. 12 and 15. Both half-sections are located, by the front and rear wing-spar fittings, on trunnion-mounted pick-up arms carried on a pedestal (Fig. 10), which can be lowered on hinges for subsequent removal of the fuselage assembly. Levelling pads are an integral part of the pick-up arms and serve

Fig. 15. Another view, from the forward end, of the fuselage on the boxing-up fixture.

Fig. 16. Mosquito fuselages in the half-shell stage of assembly, coming through in quantity. In the foreground, final stages of equipment assembly are proceeding on a completed fuselage.

line to take a plywood strip which fits flush with the fuselage skin both inside and outside. On the inside, this strip is covered again with a second strip of double the width (see Fig. 13). These strips are glued and screwed in position.

Fitting Rear Bulkhead

Installation of equipment is continued after the joining of the shells; the instrument panels in the cockpit are completed and more plumbing and

Fig. 17. The levelling and drilling jigs in position for drilling No. 6 bulkhead for the fin front pick-ups.

wiring runs are completed. At the aft end of the fuselage, bulkhead No. 7 is fitted. This unit carries the rear fin and tailplane attachment fittings, all of which are assembled to jig and are in position when the bulkhead arrives at the fuselage for assembly. Bulkhead No. 6 immediately forward of No. 7 carries the front fin attachments and the slots in the top of the fuselage to admit these fittings are first cut out by drilling the ends through from the inside and removing the wood between.

A local jig for drilling the holes for the front fin pick-ups is then placed against the rear face of bulkhead No. 6. It is located by pushing lugs through the slots in the top of the fuselage to pick up, by means of detachable pins on the front end of a jig which is itself located from the rear fin pick-ups on bulkhead No. 7. The jigs are shown in position in Fig. 17. As bulkhead

No. 7 is not finally fitted at this stage a longitudinal and transverse clinometer check is made on levelling. Bulkhead No. 6 is then drilled for the fin-attachment brackets, which are bolted up to it and project through the top skin of the fuselage. Another levelling check is made over the two sets of fittings, and on their spacing, by refitting the levelling jig to pick-up on fore and aft fin mountings.

At the extreme rear, where bulkhead No. 7 is fitted, the fuselage skin is tapered off and the bulkhead frame is similarly tapered to make a close fitting joint. Some hand fitting is necessary to achieve the required accuracy of fit. A tolerance of 0.015in. only is permitted on the lateral position of the bulkhead centre line, a very close tolerance on a wood assembly. Two internal drag struts are next fitted inside the bottom of the fuselage between the two bulkheads and the tailwheel-retraction jack is also assembled.

At the forward end of the fuselage, the canopy over the pilot's cockpit is also assembled. The edges of the opening are chamfered and fitted to a large wooden profile template or gauge, which is brought along on an overhead crane and lowered into place. After this operation has been completed, the edges of the opening are brush painted and the canopy itself, which is a sub-assembly, is lowered by overhead crane into place and bolted to the fuselage.

Final stages of installation before

the fuselage is taken to the final assembly line include the fitting of the instrument panels, radio equipment, and the completion as far as possible of hydraulic and electrical circuits, particularly in the nose section of the aircraft.

Wing Production

In appearance, the Mosquito wing is very distinctive. The sharp taper on the trailing edge, with the lesser degree of taper on the leading edge, gives it, in plan, a pointed shape characteristic of several earlier de Havilland types. It is made in one piece from tip to tip and, structurally, is based on conventional two-spar practice with the usual interspar rib structure. The stressed-skin covering, however, is a distinct departure from the usual type of construction. The birch plywood skin is reinforced by closely-spaced, square - section, spanwise stringers of Douglas fir, which, over the upper surface of the wing are sandwiched between a double covering of skin. On the under surface, the outboard panels of the skin are of identical construction but with only one skin. Over the centre portion of the span, where the fuel tanks are housed between the spars, the wing under surface is completed by stressed covers to the tank bays. These tank doors are of balsa-plywood sandwich construction, with bolting edges of shear resistant material.

Both main spars are of box construction, with laminated spruce booms and plywood webs on each side. The front spar is a straight member, which does not differ greatly from the usual component of this type. The rear spar embodies the sharp, swept forward outline of the trailing portion of the wing. Owing to the one-piece construction, the wing dihedral is combined with this forward sweep and the whole spar constitutes a most in-

Fig. 18. Bottom rear spar-boom laminations in the cementing fixture.

teresting production problem indeed.

In their finished state the spars have a length of more than 50ft., and their handling alone presents a number of problems. The rear spar booms are built up from a centre section, two outer portions and two tip extensions, all scarf-jointed with Beetle synthetic resin cement. Each section is of laminated construction.

Spar-boom Manufacture

Hitherto the top boom has been made up of three laminations, each of 1·45in. thickness, while a larger number of laminations 0·4in. in thickness have been used for the main lower boom. The heavier laminations obviously impose a higher standard as regards freedom from faults in the log from which the laminations are cut

and, in order to make the most economical use of all timber supplies available, a changeover is being made to the thinner laminations for the top spar. This modification has been made with an increase in weight of only 3lb., which represents the additional cement required to join the larger number of laminations. In the present article, the production of the original type will be described.

A considerable economy in time and a 40 per cent. saving in timber have been effected as a result of developing a special technique for sawing laminations to an accuracy of 0·010in., and so eliminating the need for a sizing operation after sawing. The slightly rougher finish left by sawing is also advantageous for the making of a cemented joint. Fourteen booms can

Fig. 19. Spindling lightening recesses in a top spar-boom extension.

Fig. 20. Rough machining a bottom boom on an overhead planer.

in this way be produced from the amount of timber that was formerly required for ten.

Somewhat different procedure is used in building up the top boom with the heavy laminations and the lower boom with its 0.4in. thicknesses. In the case of the top boom, the centre section and main outer portions are built up to full depth as separate units. Beetle cement is applied to the faces of the laminations with rubber squeegees and the assembled pack is placed in a screw-press for bonding.

Lightening recesses, increasing in depth towards the outboard ends, are machined in the rear faces of the outer

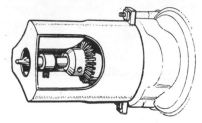

Fig. 21. Stops are fitted to the portable electric drills to govern the depth of countersinking in plywood web and skin panels.

sections of the top boom. The recesses are first marked out on the boom from templates. A cutter formed to the profile of half the recess is used for this operation, which is done on a spindling machine (Fig. 19).

The short centre section of the top boom, corresponding with the width

Fig. 22. The fixture in which the spar boom sections are scarf-jointed to form the complete unit.

of the fuselage in the aircraft, is chamfered at each end to the angle of forward sweep on the outboard portions of the boom. A direct scarf joint can, therefore, be made between the side of the outboard section and the end of the centre section, and the question of forming, a somewhat difficult matter with such heavy laminations, does not arise.

With the lower boom, however, a different method of assembly is adopted. Each lamination is made up to its full span, less tip extensions, by scarfing and cementing a number of shorter lengths. The dihedral which is "in way of face" of the laminations is formed by a specially shaped section in the centre. The pack of laminations for the complete boom is then stacked in proper sequence and each one coated on one side with Beetle cement, applied, as usual, with rubber squeegees.

As they are coated, the laminations are transferred to the bonding fixture, shown in Fig. 18, which is, in effect, a large forming block, shaped to the required angle of forward sweep. When the whole pack is in position a metal-faced pad is placed over the top one and screwclamps are swung into position to force the laminations hard down over the formed base of the fixture. In drying, the laminations take a permanent set to the shape required.

After removal from the fixture the booms are stacked on vertical racks from which they are taken for rough machining to an

overhand planer (Fig. 20). During this operation the ends of the boom are supported on rollers and it is machined down as close to finished size as is feasible, having regard to the accuracy obtainable by this method of machining. Finishing of the boom to size is done on another fixture, the small amount remaining being re-

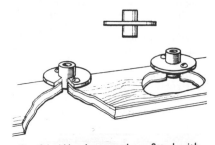

Fig. 24. Wooden templates fitted with case-hardened mild-steel bushes are used for drilling the plywood panels.

moved by hand planing and checked by straight-edges laid across datum blocks on each side.

Spar-webs

Plywood spar-webs are made up from a number of short lengths which are spindled to shape and drilled for attachment to the boom in simple wooden jigs. Skin sections are located two or three at a time between pads on a baseboard and the drill template is placed over it, drilling being performed with portable tools.

The construction of these templates is of interest. Each drill bush has a good effective length, a very desirable feature where portable drills are being used, even though the length of hole to be drilled is, as in this case, only very short. These bushes are of case-hardened mild steel with a central

Fig. 23. Scarfing spar webs from smaller sections of plywood.

flange through which they are screwed to the face of the template. The shank of the bush on one side of the flange is let into the template.

Scarfing of the lengths of plywood to make the complete web is done by machine and also by hand on a horizontal table (Fig. 23). Each section is located between pads and arranged with the scarfed ends overlapping. Beetle cement is applied and

Fig. 25. Control panel for the spar electrical heating equipment. The spar sections for thermostatic regulation of heating are on the extreme left and right.

Fig. 26. One of the heating panels in position on the spar. Connection is made to the bus-bars by a flexible lead, through a balancing resistance.

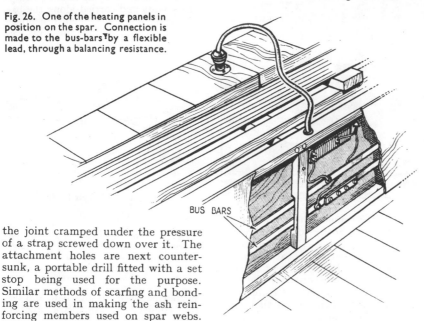

BUS BARS

the joint cramped under the pressure of a strap screwed down over it. The attachment holes are next countersunk, a portable drill fitted with a set stop being used for the purpose. Similar methods of scarfing and bonding are used in making the ash reinforcing members used on spar webs.

Spar Assembly

Assembly of the spars is done in large fixtures with sloped access platforms made necessary by the swept-forward outboard ends of the spars

(Fig. 27). The booms are very simply located between blocks on the base of the fixture at each side. Longitudinal location is given by the combined forward sweep and dihedral in conjunction with a chordwise taper on the top and bottom faces of the boom. A slight excess length is left ·on the tips of the booms and is trimmed off in the fixture. It is worthy of note that a tolerance of 0·020in. is maintained on the overall length of each half of the boom.

The booms are secured in the fixture by wedges driven in against their inner faces and the locating blocks. Spruce spacing members between the booms are positioned on the fixture to templates located over the faces of the booms. Beetle cement is applied to the booms and spacers and to the underside of the web, which is then laid in position and screwed down. the screws providing the pressure required for the bonding of the joint. Owing to the increased setting time required for this cement as compared with the adhesive formerly used, and the fact that no other work can be done on

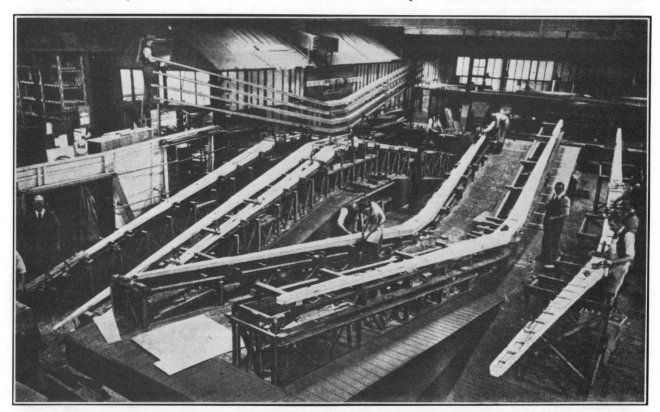

Fig. 27. The main fixtures for rear-spar assembly. For applying the skin to the forward face of the spar the inverted fixture in the foreground is used.

the top of the spar web in the fixture (Fig. 26). These panels are of wood with wire heating-elements embedded in their bottom faces and connected to a plug socket at one end of the top face. Heating current is supplied from bus-bars carrying about 40 amps. at 25 volts and carried along the base of the fixture. Balancing resistances, to adjust the heating given by each panel to the correct figure, are con-

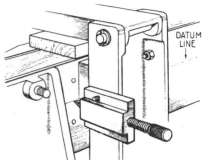

Fig. 29. The height of the spar in the drill jig is adjusted by setting a datum-line on the web face to a series of spring-loaded pointers.

nected between the bus-bars and flexible connections which are taken off at intervals to the panels.

Thermostatic Heating Control

Thermostatic control of heating is maintained in a most interesting and ingenious way. A section of spar has been built up and is mounted on the

---

the spar during the setting period, it became necessary to devise a means of accelerating the setting of the cement in order to avoid creating a bottleneck at this stage.

Accelerating Adhesive Setting

It is well known that the setting of synthetic adhesives can be speeded up by raising their temperature* and it was decided to make use of this fact and to adopt some form of electrical heating. The problem was not a simple one, as direct application of

See "Aircraft Production" for April, 1943.

heat to the cement line is not possible and the area to be considered varied considerably across the span of the spar. A further complication was introduced by the fact that a door at one side of the shop near the starboard end of the spar fixtures gave a different temperature to that existing at the port end.

With the co-operation and advice of the Northmet Car Company, a scheme was finally worked out and successfully applied. It is based on the use of a number of sectional heating panels which are placed on

Fig. 30. The complete drilling jig for the rear spar.

Fig. 31. Skin panels being drilled (left) from wooden templates. In the centre, the stringers have been located in the slots of the assembly fixture and on the right, the skin is being applied over a similar set.

control panel: the thermostat is mounted on the end of this spar section and connected directly to the cement line. As the temperature rises or falls in the shop, the thermostat cuts the circuit out or in to maintain the heating temperature within a range of $\pm 1\frac{1}{2}$ deg. F. Provision is made for plugging in thermometers on the cement line to give a direct reading of the actual temperature at any time.

When the setting of the first side web has been completed and the ash reinforcing members assembled, the spar is removed from the fixture, turned over and placed in the inverted fixture with the outboard ends raised, for the application of the front web which is done in the same way as the first. The rib posts are assembled in this fixture on the front web of the spar. A post on the centre-line of the spar is first assembled to a datum line on the web and a bar with a series of stops corresponding to the rib stations is located from it. This bar is in three sections, linked together, and corresponding with the straight centre section and the two swept-forward outboard portions. The ribs are positioned to the stops on the bar and cemented and screwed in place.

Spar Drilling

From the inverted fixture, the spars are transferred to the drilling jig, in which the holes are drilled for all the metal fittings to be mounted on the spar, such as the undercarriage and the hinge brackets for the ailerons on the rear face. The jig is similar in principle to that usually employed for the drilling of wood aircraft spars,* and is shown in Figs. 28 and 30. Bush

plates are provided for drilling from both faces of the spar, which is supported and located between each pair of wooden pads below and on each side. The bottom pad can be raised or lowered by screw adjustment to set the height of the spar in the jig and the pads on one side of the fixture can be screwed in to clamp the spar in position. Setting is made to a longitudinal datum line and a vertical datum at the centre line, both of which are laid out on the spar web from templates in the main assembly fixture. The longitudinal datum is set to pointers mounted on the jig base (Fig. 29).

A colour code is used to distinguish the bushes for different drill diameters and a portable drill is used. An interesting point is that a dead stop is fitted at each end of the jig, which imposes a very definite check on the overall length of the spar. The production of these large components in wood to a tolerance of $\pm$ 0·040in. over a length of 50ft. is an achievement which reflects the greatest credit upon the sub-contractors.

Skin Panels

It will be remembered from the earlier description that the skin is reinforced by square section stringers of Douglas fir, and that there is a "sandwich" of stringers between two skins on the top surface of the wing. In the case of the outboard lower-surface panels and of the inner skins of the top surface, the stringers are assembled with the skin as a single unit. The inner top skin with its stringers is known as the inner shell and is made in three sections, a narrow-centre portion with the stringers projecting on each side and scarfed at the ends ready for jointing with those of the two outboard sections.

As in the case of the boom laminations, the stringers are made up by scarfing from shorter lengths and the taper of the scarfing itself is saw cut. The skins, like the spar webs, are assembled by scarfing from smaller sections and are drilled from large wooden templates similar to those used for the spar webs. Procedure is, in fact, exactly similar, but on a larger scale. Countersinking of the holes is done in the same way. Shaping of the skins is done by running a portable electric saw along the edges of the templates located on the skin from the drilled holes.

Assembly Methods

For their assembly with the skin the stringers are located in metal-faced slots on a horizontal table (Fig. 31). Cement is applied to the top faces and to the underside of the skin, which is then laid over it, located from the edges by metal stops on the table and screwed down by pump screwdrivers. The screws supply the necessary pressure to make the joint. Stringers and skin for the centre section of the inner shell are assembled in a vertical fixture, otherwise procedure is the same.

After the assembly of reinforcing plies the cutting out of access, inspection and other apertures, the skins are ready for doping and painting. Red dope is sprayed over the outer surfaces while the stringers and inner skin are coated with white water-resisting cellulose paint. A masking frame is used to keep the outer faces of the stringers free from paint for the cement joint made later with the outer skin. The skins are then ready for assembly with the wing. The outer skin for the upper surface is made up into a single unit by scarfing and cementing smaller sections in the usual way.

Mosquito Versatility

The fighter version of the Mosquito with auxiliary long-range drop tanks beneath the wings. The 0.303in. machine guns in the nose and position of the 20mm. cannon beneath are clearly seen.

Fifteen Functions Performed by Different Versions : Speed Always the Vital Factor : Long-range and Night-fighter Types

TIME was when the R.A.F. favoured the policy of a highly specialised aircraft for each specific duty, believing that any attempt at versatility in a military type must inevitably produce a Jack-of-all-trades-master-of-none result, and in the days when almost all designs incorporated "built-in headwinds" there was a lot to be said for this argument.

If you wanted a fighter, then speed was the *sine qua non*. And since any attempt to broaden its field of usefulness could only be at the expense of speed, then it had to remain a fighter and nothing more. It could be inferred from this that speed was not then considered to be the first essential of such types as bombers, observation and reconnaissance aircraft; other considerations came first, according to the actual duty to be performed, after which as much speed as was possible remained a desirable secondary quality.

But experience, particularly during the present war, has changed all that. More and more it came to be realised that with very few exceptions indeed speed was of paramount importance in almost every kind of military aircraft, and it was found that the fighter type, suitably modified, could fulfil a variety of duties far more effectively than many machines specifically designed to perform them. Thus we saw the slow, be-slotted and be-flapped Lysander give place to the sleek, swift Mustang for reconnaissance duties with Army Co-operation squadrons. Not that this is any reflection on the redoubtable "Lizzie"; it was designed to fulfil certain clearly laid-down requirements and it did so admirably; it is merely that the requirements have changed in the light of steadily accumulating experience. Later still, the Spitfire's speed has been found a life-saving factor in the Air/Sea Rescue Service.

A Brilliant Example

To-day one could scarcely have a more brilliant example of versatility plus really high speed than is provided by the de Havilland Mosquito, and it is no undue flattery to say that its designers—specialists in civil aircraft producing their first military type of the present war—were well to the fore in appreciating the true value of speed in all military flying operations. Indeed, one can go back to 1934 and the D.H. Comet, designed for the England-Australia race, for the first intimation of the trend of de Havilland thought in this direction. For the potentialities of the Comet, acknowledged forebear of the Mosquito, were voiced by Capt. Hubert Broad, at that time de Havilland's chief test pilot, who expressed the opinion that with a few modifications this long-distance race winner could be converted into a very useful bomber. That it already possessed the chief attribute of a fighter was self-evident.

It is now permitted to enumerate no fewer than 15 different variations of the Mosquito, eight functioning by night and seven by day, and the latest to be released for publication, that of night-fighter, provokes an interesting reflection. It is that the Germans have no adequate answer to the Mosquito as a night-bomber, nor as a night-intruder, but the Mosquito is proving one of the most effective night-fighters to deal with the nuisance raids by German fighter-bombers.

SWIFT BUT CIVIL : The normal-range transport Mosquito in service with British Overseas Airways. It can safely be described as the fastest civil aircraft in the world.

MOSQUITO VERSATILITY

Another duty being most capably fulfilled by the Mosquito is that of photographic reconnaissance, as described in *Flight* of Sept. 9th, and the P.R.U. employs two variants of the type, one for low and medium altitude photography (the former being known, with a certain grim humour, as "dicing") and another for high altitude work, which implies the use of Merlin LXI engines with two-stage, two-speed superchargers.

Yet another version of the Mosquito, the existence of which has only recently been officially disclosed, is the civil type in the service of British Overseas Airways Corporation. Strictly speaking, there are two editions of the civil, or transport model, one operating by day and the other by night, and it will be fresh in every reader's mind that the pilot and navigator of one of these civil—

HIGH, WIDE AND HANDSOME : The leading-edge radiators are apparent in this front view of the high-altitude, long-range Mosquito used by the Photographic Reconnaissance Unit. Note the wing drop-tanks.

TIGHT TURN : Manœuvrability is demonstrated in this action picture of a normal-range Mosquito fighter.

and therefore unarmed—Mosquitoes have been awarded the O.B.E. and M.B.E. respectively for their courage and devotion to duty over a considerable period in regularly flying between this country and Sweden in defiance of enemy fighters.

It will perhaps now be as well to tabulate the fifteen different variations of the Mosquito, as follows:—

Fighter.
    (*a*) By day, long-range oceanic patrol, *e.g.*, Bay of Biscay.
    (*b*) By night, (i) Home defence interceptor.
               (ii) Differently equipped, night fighter over enemy territory.
Intruder.
    (*a*) Day intruder.   (*b*) Night intruder.
Bomber.
    (*a*) By day, (i) Low attack. (ii) High attack.
    (*b*) By night, (i) Low attack. (ii) High attack.
Photographic Reconnaissance.
    (*a*) By day, (i) Low and medium altitude.
               (ii) High altitude.
    (*b*) By night (differently equipped):
               (i) Low and medium altitude.
               (ii) High altitude.
Transport.
    (*a*) By day.   (*b*) By night.

Photographs have previously been published showing long-range drop-tanks beneath the wings of bomber, P.R.U., and civil versions, but to-day we include one showing them fitted to the Mark II fighter type also.

An R.A.A.F. night-fighter squadron in Britain was among the first units to use Mosquitoes for this work, and bagged its first raider on the night of May 7th last. Since then it has made more than 400 sorties, including daylight train-busting excursions, at remarkably low cost to itself. But perhaps the most eloquent testimony to the value of speed is Bomber Command's record, for during the past six months only 11 Mosquitoes have failed to return from some 60 major raids on Germany and nearly 1,000 other sorties against the enemy's industrial centres.

LONG-DISTANCE NIGHT EXPRESS : The long-range version of the civil Mosquito in service with British Overseas Airways, with flaps partially lowered, faces up the flarepath in readiness for a nocturnal take-off.

MOSQUITO
Fastest Aircraft in Service in the World

Another triumph by

DE HAVILLAND
Leading Builders of Transport Aircraft in the British Empire

A UNIT OF THE
DE HAVILLAND
WORLD FORMATION

de Havilland Aircraft   de Havilland Engines
de Havilland Propellers   Components and
Accessories   Light Alloy Engineering
Flying Training   Technical Education

In the attack today — on the trade routes of the future

THE WAR IN THE AIR

NIGHT FIGHTER.—de Havilland Mosquito II two-seat fighters, powered with Rolls-Royce Merlin 21 liquid-cooled upright-vee in-line motors, have been in operational service with R.A.F. Fighter Command for some time past. Already they have a considerable bag of victories, particularly against German high-speed nuisance night raiders.

LONG-RANGE DAY BOMBERS.—Many de Havilland Mosquito two-motor day fighters (two Rolls-Royce Merlin 21 motors) are now fitted with long-range fuel tanks slung externally beneath the wings outboard of the motor nacelles. Armament is concentrated in the nose.

AIRCRAFT RECOGNITION

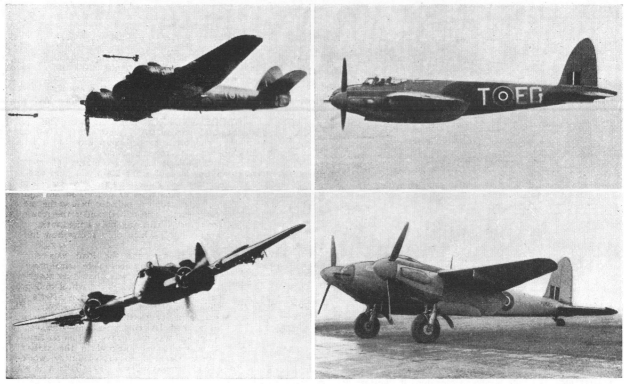

RECOGNITION POINTS.—The above photographs and the tone drawings on the opposite page should aid in the swift identification of (left) the R.P. Beaufighter and (right) the F.B. Mosquito.

THE BRISTOL BEAU-FIGHTER R.P.X (two 1,650 h.p. Bristol Hercules XVI 14-cylinder two-row air-cooled radial motors) and the de Havilland Mosquito fighter - bomber (two Rolls-Royce Merlin XX series 12-cylinder upright vee liquid-cooled motors) were the subjects of the previous identification tests.

Rocket projectiles and torpedoes have almost entirely replaced bombs in the armoury of Coastal Command, and it is Bristol Beaufighters that are being employed to carry these weapons against the enemy. R.P. Beaufighters are operating in large numbers, both with Coastal Command in Home waters and the Mediterranean Allied Coastal Air Force in the Northern Mediterranean. Rocket projectiles have no recoil like that of an ordinary gun and have very flat trajectories, making their installation in batteries on an aeroplane a simple matter. A large flanged tray under each wing provides the attachment for the eight rocket rails and the electric cables for the firing mechanism are led from the undersurface of the wing to the back of each rocket.

TWO MORE TESTS.—Photographs, drawings and descriptions of the above aeroplanes will be published at a future date.

The outline of the R.P. Beaufighter differs only slightly from that of earlier versions, the chief point to note being the rails under the wings. D.H. Mosquitoes are now fast becoming as complex in their variations as the Spitfire, but lack of mark number releases makes the task of sorting them out a hard one.

The Mosquito dealt with in this feature is the standard fighter - bomber version which carries two 500-lb. bombs in the rear of the bomb bay and one 500-lb. bomb under each wing on a detachable rack. Combined with an armament of four 20 mm. cannon and four 0.303-in. machine-guns and a very high speed, these make the F.B. Mosquito one of the most formidable "transport-strafers" in the World. Mosquito bombers also have provision for the bombs under the wings, giving a maximum load of six 500-lb. bombs, although the latest Mosquito now carries one 4,000-lb. "cookie." The bomb racks under the wings can be replaced by fuel tanks.

"Invasion" stripes have been omitted on the tone drawings to preserve clarity of outline.

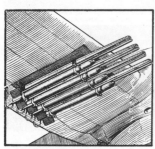  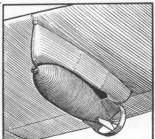 

CHARACTERISTIC FEATURES.—(a) Rocket installation on the Beaufighter. (b) Tail of the Beaufighter. (c) Bomb rack and 500-lb. bomb on Mosquito port wing. (d) Tail of Mosquito.

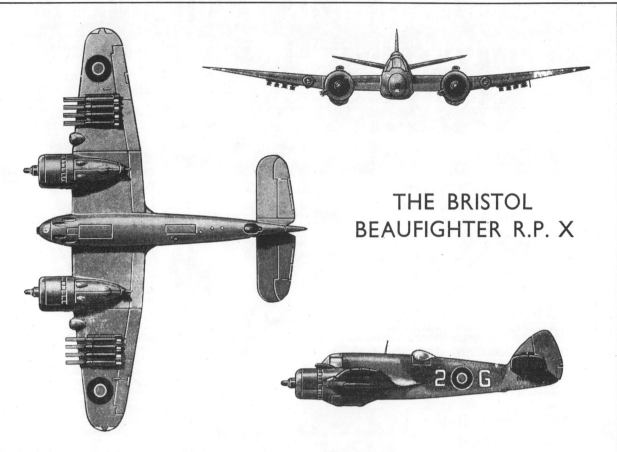

THE BRISTOL
BEAUFIGHTER R.P. X

Span, 57 ft. 10 ins. ; length, 41 ft. 4 ins.

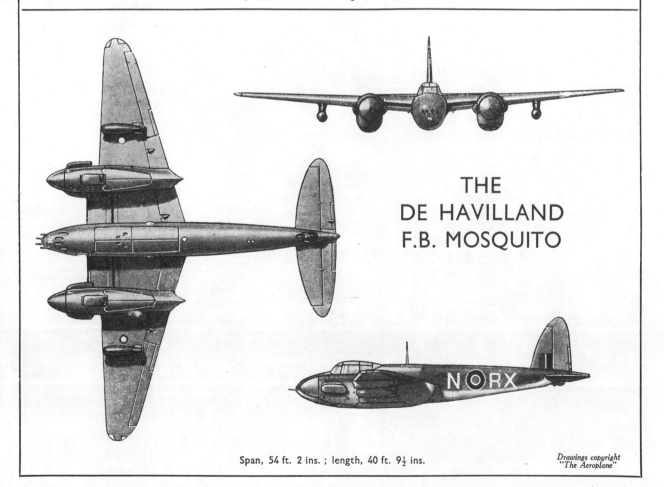

THE
DE HAVILLAND
F.B. MOSQUITO

Span, 54 ft. 2 ins. ; length, 40 ft. 9½ ins.

From the Australian de Havilland factories MOSQUITOES fly to the attack!

The de Havilland war effort is strategically world dispersed. War found an established chain of factories, forged from 1927 onward for the service of civil flying, ready to expand and turn over to combat aircraft.

Mosquito production in Britain and Canada is now augmented by a mounting flow of fighter-bombers from the Australian plants.

de Havilland propellers and Gipsy engines also are manufactured in the Commonwealth.

When war is won the Empire wide de Havilland enterprise will be ready again to serve the needs of peace.

DE HAVILLAND
AIRCRAFT · ENGINES · PROPELLERS

MOSQUITO *Fighter-Bombers*

The Work of Our Fastest Twin-engined Fighter-Bombers Which Harass the Germans by Day and Night

From Our Own Correspondent at S.H.A.E.F.

MANY of the prisoners captured in the Falaise pocket and in subsequent operations have said that our night fighter-bomber patrols were the worst part of all the fighting. This is easily understandable if one goes to one of the stations which operate these patrols and sees the quality of work put in to make the patrols as effective as possible. Hundreds are flown every night in all weathers, and much useful work is done.

It is impossible to go on ops. with these boys because the Mosquito only carries two—pilot and navigator—and there is no place at all where a third person could be tucked away. However, having flown in one of the Mosquitoes during a night-flying test, and having had a very detailed description of a typical operation from a wing commander who has done many of them, I think I can reconstruct the picture fairly faithfully.

Aircrews who have flown the night before generally put in an appearance again round about 2 p.m. in readiness for another night's work. These boys fly quite a number of ops. in a short while—two in a night is quite common. When some very special target, such as the Falaise pocket, is being attacked, as many as three trips to the battle area are made during the hours of darkness. This is a very great tribute to the aircrews, to the ground crews

Ground crews of a Polish squadron prepare one of their Mosquitoes for operations.

for a quick turn-round, and to the Mosquito for its speed.

During the early evening the night's work for the Wing comes in from Group H.Q., and a conference is held to allocate the tasks. This conference is attended by the Station Commander, Group Capt. L. W. C. Bower, D.F.C., Wing Cdr. Intelligence, Wing Cdr. Flying, an Air Liaison Officer and the squadron commanders. In addition to these, specialist officers—armourer, engineer, meteorological and flying control—will attend to obtain a clear picture of the night's work.

According to the distance away of the target areas, orders will be given for long-range tanks to be fitted or removed. The armament officer will decide what load is to be carried and the types of bombs, flares and illuminating cartridges to be used. "Met." gives a picture of what the weather will be, both at home and over the target, during the

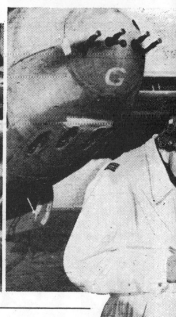

MOSQUITO FIGHTER-BOMBERS

whole night. Flying Control decides which runway will be used and the compass bearing of that runway. Later the individual crews are briefed. Pilots and navigators assemble in the squadron crewroom, and the squadron commander reads out the names of the crews and chalks on the board their take-off time and time over the target.

Next, the actual routes of the patrols are given and also the heights and routes by which the Mosquitoes fly to and return from the target areas. The Air Liaison Officer then gives a general picture of the progress of the battle, with special stress on the effect of the patrols which are just going to be flown. On the night I was there the crews were given the railways in the Lille, Cambrai, Montdidier, Amiens area to watch for trains taking supplies to the battle area and taking flying bombs to the launching sites. Many of the crews reported a blank, but a train, believed to be carrying flying bombs, was attacked and destroyed. Even the negative reconnaissance is useful because it fre-

A low-level attack with cannon, machine-guns and bombs on a train. This photograph, taken through the bullet-proof windscreen, gives the navigator's view of the attack.

(Left.) Pilots and navigators in the crew room waiting for briefing to begin. (Centre.) A pilot signs for his Mosquito before getting on board. (Right.) A pilot and his navigator seated in the cockpit. The reflector sight projects from the dashboard.

quently ties up with earlier operations farther to the rear and shows their effectiveness.

Finally, the Intelligence Officer gives details of flak patches and of such friendly aircraft as will be working in the vicinity of the patrols. For the old hands this is the end of the briefing. The squadron commander, however, takes new crews on one side and, out of the wealth of his experience, tells them how to do the job most effectively at the least possible risk to themselves. He goes over all the briefing again, patiently enlarging on all the salient points—the effect of autumn mist on flares—other airfields at which to land if home base is foggy—how to tackle a train with flak wagons—the evasion of flak and search-lights—enemy night-fighter tactics when bombing under parachute flares. He reminds them to watch length of patrol and petrol consumption carefully, and finally tells them not to be greedy over any "joy" they might find but to make their attack and get out of the way after calling up others.

Out on the airfield the air crews make a thorough check of their aircraft. Bombs, guns, flares and anything else which has been fitted since the night-flying test was carried out, are carefully gone over.

As each take-off time approaches, the pilot and navigator concerned leave their midnight operational breakfast of bacon, eggs and chips and go to their Mosquito at the dispersal point. Flying gear is put on and they settle comfortably in the restricted but not cramped cockpit. Oxygen and V.H.F. radio are

connected, and the oxygen is turned on to the equivalent of 10,000-15,000ft. This has been found a distinct help in keeping wakeful and helping night vision. The Mosquito crews carry a large supply and use it the whole time at whatever height they are flying.

Start-up

The Rolls-Royce Merlin XXIs are then started up—starboard first, then port—and left running at 1,200 r.p.m. While they are warming up, the trimming tabs are set. Elevator three-quarter division forward, rudder three-quarter division starboard, ailerons neutral. V.H.F. radio is switched on and altimeter set to read zero. With the engines warmed up to 40-50 degrees centigrade the radiator shutters are opened and the chocks signalled away by the flicking on and off of the navigation lights.

The confidence trick. Fly at 1,000ft. with one engine of the Mosquito stopped and the airscrew feathered.

MOSQUITO FIGHTER-BOMBERS

Before the aircraft starts to move, Flying Control is called up on the radio and asked for permission to taxy. As the Q.K. to taxy comes through the earphones one can see mentally the control officer take the stud with the aircraft's number on it from a panel marked " dispersals " and stick it in a rectangle marked " taxying."

Dispersal is frequently some way from the marshalling point, and as the aircraft taxies gently round the perimeter track there is time to listen to Control talking to other machines either in the air or on the ground. A picture is formed in the crews' minds of what other traffic is about. Another report to Control is made when the blue lights of the marshalling point are reached, and in Control the stud moves on again from the taxying square. Windows are closed and permission asked and given to move on to runway. The Control stud shifts again. With the present-day close cowling of engines, priority is given to aircraft waiting to take off because of possible overheating.

On the runway the Rolls engines are blipped up to 2,500

For distant operations. Screwing up the central fitting of a long-range tank under the starboard wing.

r.p.m. just to clear the plugs. There is no need to run them up. The reliability is such that if they were working perfectly at the time of the previous landing they are taken on trust to be still all right.

Friction throttle-control dampers are tightened, gyro compass is set to the bearing of the runway (this helps the pilot to keep from swinging during take-off on a bad night), the two engines are set to same speed, and the wing flaps lowered 12 to 15 degrees. All set ; Control is called again and permission to " scramble " obtained. Brakes are released and throttles opened up—with a slight lead on the port engine to counteract any tendency to swing—until 3,000 r.p.m. is reached with plus 9 lb. reading on the boost gauge. A.S.I. showing 110-120 m.p.h., a slight easing back of the stick, and the machine is off. At this stage the pilot changes hands on the stick and selects " undercarriage up," but meanwhile the Mosquito is climbing only a very little for the next 20 or 30 seconds until 170 m.p.h., the minimum single-engine speed, is reached. At 500ft. the flaps are raised and the resulting nose-down attitude re-trimmed.

On Patrol

After climb to patrol height the Mosquito orbits until " set course " time arrives—usually one or two circuits is ample—then the radiator shutters are closed and course set with about 250 m.p.h. showing on the A.S.I., 2,400 r.p.m. on the rev counter and plus $4\frac{1}{2}$ lb. boost.

By now the eyes have become " dark accustomed " and the visual purple in them is at its maximum. Below, in all directions, can be seen coloured lights, flashing beacons, stationary searchlights and other signs which, to those who understand, represent an air force at war at night. Straight ahead the horizon can be sensed rather than seen. It is where the stars become dim and stop and nothing begins.

As the coast is passed the speed is stepped up considerably and the sky is searched continuously for enemy night fighters. A prominent feature of the enemy landscape from which to start work is pinpointed, and as the patrol area is approached other crews are heard calling Mosquitoes in the vicinity to share in any " joy " they have found. A number of lights is seen, but it is suspected that these are mostly French. The mentality of the Frenchman is such that it is impossible for him to keep a strict blackout. Down below, the railways are deserted, but a row of dim lights appears on a road. By the experienced Mosquito crew this is immediately interpreted as a road convoy and probably a very urgent one to be moving regardless of air attack. Bomb doors are opened and a cluster of flares dropped. Five seconds later a radius of half a mile is lit up, showing a road through a wood with vehicles passing along it. A diving attack is made with short bursts of fire from the four 20 mm. cannon and four .303in. machine guns until the bombs are released at the bottom of the dive. To avoid any flak or night fighters, evasive action is taken during which the impact of the

Three stages in the daylight attack by Mosquitoes on the German barracks in a technical school at Egletons.

(Top left.) Stripping covers ready for flying. With the exception of cockpit and engine covers, the Mosquitoes stand out in all weathers with no protection. (Top right.) Bombing-up the port wing rack with a 500-lb. bomb by the aid of a pair of "hockey sticks." (Centre left.) An armourer re-arms the four machine-guns in the nose. (Centre right.) Filling the oxygen bottles in the fuselage. A six-hour supply is carried. (Bottom left.) The telephone exchange in the mobile control unit.

blast of the bombs is felt. A small fire is seen, and another cannon attack is made; the resulting fires show the target to be worthwhile. Other Mosquitoes are called *en clair* to join in the fun. While this is going on the flares lose height almost imperceptibly. In this very modern and complicated air war the earliest known method of flying is called in to help. The hot air rising from the flares supplies lift to the parachutes which support them. The Montgolfiers used this principle in their first successful ascents in hot-air balloons.

On the way home a wise pilot flies gently. While engine damage is self-evident, it is impossible to know what flak damage the airframe has sustained.

As Britain's coast is crossed again, that lovely feeling of being back home is felt. Searchlights and shadows are no longer sinister. Control is called and, after it has been given the aircraft's call sign, is told "feet are dry" to let them know that the machine is over the Channel safely. Barometric pressure is given to reset the altimeter. Height at which to orbit the airfield is also given, and while waiting for instructions to "pancake" a picture of the aircraft in the locality can again be obtained by listening to the Control Officer despatching and landing them.

MOSQUITO FIGHTER-BOMBERS

(Below.) In field conditions, airfield control is carried out from a mobile unit. While it is not as convenient as the static control room, a considerable volume of traffic can be handled.

(Above.) A Control Officer in the conning tower of the mobile unit. He is talking to aircraft at the dispersal points which are asking for permission to taxy.

When next in turn to land, speed is reduced to 180 m.p.h. and the undercarriage lowered. This is checked twice to make certain that the tail wheel is down. Flaps are put down 30 deg. during a cross-wind approach, with speed down to 150 m.p.h., and fully down as a final turn is made into the runway funnels. Engines are revved up to 2,850 r.p.m. in fine pitch and throttled right back as the coloured lights of the glide-path indicator are approached. A slight scrape is heard as the touch-down is made at about 115 m.p.h. because the wheels refuse to accelerate instantaneously. A series of hisses occur as the compressed air is used to operate the brakes to avoid swing and to pull up the Mosquito. It comes to a standstill and then taxies off the runway. Immediately Control is called and "runway clear" is given, as well as "good night," plus any little pleasantry appropriate to the moment.

At the dispersal point the oxygen and V.H.F. radio is switched off and the engines stopped by the aid of the cut-outs. The ground crew are eager to know if everything was all right and to hear of the night's work.

Interrogation

Transport takes the crew to the briefing room, where the flight will be discussed in detail. Down each side of the marquee are tables and chairs, cigarettes and hot tea with a lacing of rum. Each table has an Interrogating Officer, and a crew coming in goes to any vacant table. At one end of the tables is another table at which the chief Intelligence Officer sits. All the reports are passed to him and co-ordinated by him into a comprehensive report to Group. If there is something requiring urgent attention, Group is, of course, told straightway. These interrogating officers need to be good psychologists, and from just this point of view it is interesting to watch them at work. Some crews have to have every comment dragged out of them by leading questions; others "shoot an imperial line" and have to be curbed. The most difficult of all are those who say that nothing very much has happened and then just casually mention some terrific job of work they have done.

It must not be thought that all the flying is done at night. Actually this is quite a recent innovation, and low-level attacks by day are still being made. It was an inspiration to hear a French crew describe a recent attack on a suspected Gestapo H.Q. They were the only French crew in the Wing. René was an Armée de l'air pilot of 15 years' experience, and Jacque, his navigator, a student in Paris. The target was at Egletons, 50 miles south-east of Limoges. In it were a number of Germans surrounded by the Maquis. Bombing had to be accurate because the Maquis were within rifle-shot distance. According to these two French lads, it was *very* accurate. They saw the bombs from the Mosquito in front of them go straight in through the front door and, knowing that their own bombs were in formation with them, presumed that they entered close by. Out of 14 aircraft one was lost by a stray machine-gun bullet. Shortly afterwards came the message from the Maquis, "Many thanks for magnificent bombing."

The other week this Wing celebrated their 2,000th sortie since D-Day. This has meant very hard work by the aircrews; often two, and occasionally three, journeys were made in a night. It is, however, an equal tribute to the ground crews and aircraft that such pressure can be maintained. It would not be possible at all if the majority of the flying had to be done by day. As it is; the Mosquitoes fly by night and are serviced by day. Although the flying hours are 200 per cent. over the pre-invasion figures, yet serviceability is still well over 95 per cent. of the aircraft on charge. The wooden airframes stand up to all weathers with no cover, in a manner quite equal to metal aircraft, and their only drawback is the time taken to do a glued repair.

A "BLACKBURN" SCHOLARSHIP

MR. ROBERT BLACKBURN, chairman of Blackburn Aircraft, Ltd., has undertaken to award a £100 annual scholarship to the apprentice who achieves the best results in the examinations for the ordinary National Certificate in mechanical engineering.

This offer is only open to those who take the course for the diploma in aeronautics at University College, Hull, and the first award will be made this year on the results of the examinations already held. Trade and engineer apprentices at any of the company's works are eligible.

This generous offer by Mr. Blackburn is further evidence of the increasing interest taken by the industry in the technical training of youth. The Society of British Aircraft Constructors has had in being for many years a scheme for the encouragement of boys whose parents are in financial circumstances that would otherwise debar them from following their chosen career.

Recently the de Havilland company announced details of a comprehensive education scheme for youth aeronautical training, whilst Westland Aircraft have a scheme leading ultimately to a university course at Cambridge.

Q.F. Mosquito

Coastal Command's Mark XVIII Attacks U-boats with Six-pounder Gun

ON a bleak morning last winter a homecoming U-boat was proceeding slowly to base in the calm waters south of Brest. For weeks it had been at sea, always fearing attack from constantly patrolling R.A.F. Coastal Command aircraft or from the warships of the Royal Navy.

At last they were in safe waters. . . .

But as first light stole across the sky their peace was shattered by the sounding of the alarm, "Enemy aircraft approaching."

From out of the rising sun Germans on the conning tower picked out a Mosquito aircraft diving towards them. A red ball came streaking across the sky and there was a tremendous crash as the U-boat was hit by a large shell which killed several of the crew.

This was the first attack on a U-boat, made on November 7th, 1943, with one of Britain's secret weapons—then newly introduced—the six-pounder gun carried by the Mosquito XVIII of R.A.F. Coastal Command, some details of which are now officially revealed.

Subsequent events showed that the German Admiralty had to change their in-shore tactics. Waters close to the French coast could no longer be considered safe for unescorted U-boats, and even boom- and gun-defended harbours could be penetrated by the speedy and manœuvrable Mosquito armed with the new gun.

Following the first two attacks with this weapon, the German Admiralty were forced to provide an escort of surface ships and fighters to protect their U-boats leaving or going into the harbour, thus diverting *flak* ships and armed trawlers from other duties. But in spite of *flak* from escort vessels and "umbrellas" of fighters, Coastal Command Mosquitos have persisted in their attacks on the U-boats with marked success. The accuracy of the gun was proved when a shell from it shot an enemy fighter out of the sky.

(Above) The Mosquito XVIII, with a six-pounder mounted below the four 0·303 in. machine guns in the nose.

(Right) This close-up of the "Mossie's" latest armament shows its very compact, yet accessible, installation.

The gun is slung beneath the fuselage of the Mosquito, and fires shells in quick succession as the aircraft dives to attack. Since D-day it has been used extensively aganst U-boats attempting to slink into the Channel to interfere with our landing craft. In addition to its six-pounder, the Mosquito also carries four machine guns.

The gun was first fitted to three Mosquitos which became a detachment of a secret experimental squadron. Sqn. Ldr. C. F. Rose, D.F.C., D.F.M., was the first C.O. of the detachment, which operated from an airfield in the southwest of England, and there were three other pilots— 26-year-old F/O. D. J. Turner, D.F.C., of Hornchurch, Essex; F/O. A.L. Bennett, 24-year-old R.C.A.F. pilot of Vancouver, and F/O. A. H. Hilliard, aged 24, of St. Albans, Herts.

"We were naturally very excited at being the first to use the gun," said F/O. Turner, "and we practised at sea for some time, firing one shell and then trying to hit it with others. We soon became quite accurate."

Demonstration panel for the various electrical equipment and circuits. When not in use the panel is stacked flat against the wall.

Mosquito School

R.A.F. Technical Training Command Manufacturer's Course

SHORTLY after the outbreak of war, R.A.F. Technical Training Command were faced with the acute problem of providing training for a huge and increasing personnel in the handling, repair and maintenance of aircraft and all the diverse equipment ancillary to them.

The aid of the manufacturers was invoked for the purpose of instituting at the factories instructional courses which selected personnel could attend. The venture was a great success, and the extent of the scheme was widened as more and more concerns came within the orbit of M.A.P. These manufacturers' Courses have done and are doing an excellent job, and with the advent of the Mosquito, the De Havilland Aircraft Co., Ltd., were asked to start an instructional school on similar lines to those already in existence at other firms.

When the school was started in the autumn of 1941, the average intake of students was four or five each week, and to date no fewer than 2,500 students of all nationalities have passed through. There are two courses available, the air-frame course of one week, with an average intake of 36 students, and the repair course, which lasts 14 days, with an average intake of 16 students.

Personnel from the R.A.F., W.A.A.F., B.O.A.C., A.T.A., and Fleet Air Arm have been instructed on the intricacies of the Mosquito, as well as men from the Allied Forces, Czechs, Poles, Belgians, Norwegians, Yugoslavs and Americans.

Great use is made of sectionalised components and mock-ups as a means of physical demonstration to amplify

Instruction on operation and maintenance of the undercarriage is facilitated by this full-size, powered mock-up which was constructed entirely from salvaged parts.

MOSQUITO
SCHOOL

the classroom lectures, and it is worthy of mention that every single component and unit used for demonstration in the school is salvage—nothing was abstracted from the production lines to furnish the school with demonstration equipment.

As may be seen from the accompanying illustrations, the mock-ups and demonstration panels are excellently arranged to afford students the very least difficulty in grasping the principles and essentials of the various services. In each case where motion is involved, for example the undercarriage, pneumatic and hydraulic

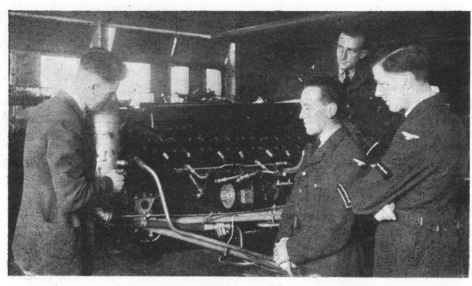

A Merlin engine is used for demonstrating the various intricacies of the power unit installation and accessories operation.

systems, etc., the layout panels and mock-ups are supplied with a power source so that operative actuation can be clearly shown.

The electrical services demonstration panel is particularly ingenious, the various control panels, components and their respective wiring circuits being positioned appropriately on a board shaped approximately in the aircraft's plan-form. This allows the students to walk around and follow the several circuits involved and, at the same time, gives a graphic representation of the particular dispositions. The demonstration panel is arranged on runners with counter-weights so that, when not in use, it can be swung into a vertical position flat against the wall and so save space.

Seeing Them Made

Classroom instruction is facilitated by the use of large-scale diagrams mounted on each side of easels fitted with casters, so enabling them to be wheeled about as required. In addition to the instruction given at the school, students are taken on conducted tours of the main works and the various dispersed factories, where they have the opportunity of seeing at first hand the various aspects of the aircraft under construction.

From experience it has been found that instructional time is lost by the necessity of students making their notes during lectures, and that mistakes inevitably occur due to the student attempting to divide his attention between his notebook and the lecturer. To minimise confusion the salient points to be dealt with during the course are arranged to coincide with the syllabus lecture notes given to each student, and as a further aid, the back of each page is left blank for any additional notes or sketches the student may wish to make. The lecture notes provided contain many diagrams which save the student time in making them personally as well as precluding the possibility of his making mistakes in drawings or annotation.

Billeting and messing for non-commissioned personnel are arranged as part of the school, but officers have to make their own arrangements locally. An excellent feature of this particular school is that both the De Havilland Senior Staff Club and the Works Social and Sports Club extend honorary membership to students for the duration of their course.

The school is run by the De Havilland Service department under the supervision of Mr. A. J. Brant, with Mr. D. W. Richardson as chief instructor. Mr. Richardson is assisted by two civilian instructors and three senior N.C.O. instructors of R.A.F. Technical Training Command.

There can be no doubt as to the value of the work the school is doing in keeping our fleets of Mosquitoes at maximum operational serviceability.

ANTI-G SUIT FOR FIGHTER PILOTS

A SIMPLE but effective means of protecting the fighter pilot from the effects of excessive "positive g" during violent manœuvres is now in general use with U.S.A.A.F fighter pilots, and was demonstrated to *Flight* last week by Capt. Donald W. Johnson, a Mustang pilot of 339 Fighter Group.

As most readers will already know, the centrifugal force set up by quick changes of direction at high speed, such as when pulling up sharply out of a dive, or making a tight turn at high speed, virtually results in the pilot's weight being temporarily increased. So long as the increase is not more than, say, 2½ to 3 times gravity, the fit man feels no ill-effects, but the performance of the modern high-speed fighter aircraft is such that, in combat, the pilot may have to perform manœuvres which will impose a load of perhaps six times gravity—6g or even more.. This causes the blood to be drained away from his head into the lower parts of his body and the almost immediate result is a black-out; not a good thing during a mix-up with a "one-ninety."

The problem, then, was to prevent this drainage of blood from the head, and it was discovered that this could readily be done by exerting a balancing pressure on the lower parts of the body.

Experiments were carried out at Wright Field (the Ameri-

can equivalent of Farnborough) and the final result was the Berger G-2 suit now in use. This is not really a suit in the accepted sense of the word, but an arrangement of five bladders, linked by straps, which fit over the pilot's stomach, thighs and calves and is worn inside the normal flying suit.

The five bladders are connected by an air-line to the exhaust side of the engine-driven vacuum pump, via a spring-valve which is automatically opened as g is increased. The action is progressive and thus the amount of pressure admitted to the bladders and thereby applied to the lower parts of the pilot's body and legs is automatically adjusted to the degree of g being exerted upon him. In effect, the same g force which is trying to drain the blood away from his head is made to operate in the opposite direction through the bladders and thus maintain a balance; so far as the pilot's physical properties are concerned, the g cancels itself out.

"We have been using this suit for some little time," Capt. Johnson told *Flight*, "and it does the job perfectly. No matter how tight the turn or how quick the pull-up from a dive, the pilot stays perfectly normal. This gives him a double advantage in that not only is he safe from blacking-out, but he does not have to bother about trying to ward off a black-out and so can look around and give his whole attention to the job in hand."

MOSQUITOES of the 8th U.S.A.A.F.

Americans Pleased with Their "Hot Ships" : Watching Weather for the Forts and Libs

By JOHN YOXALL

THE station from which the 8th U.S.A.A.F. operate their Mosquitoes is one which would please Gen. Eisenhower and his Deputy-Commander, Air Chief Marshal Sir Arthur Tedder. They are always most emphatic in their demand for co-operation between the Americans and ourselves, and no other station shows this feature to greater advantage. One sees Mosquitoes being serviced alongside Fortresses on the tarmac of a peacetime R.A.F. airfield, which was handed over lock, stock and barrel to the 8th U.S.A.A.F. The erstwhile R.A.F. officers' mess is now the American officers' club, and poker with money on the table replaces bridge with no money on the table. There is a small covey of pilots who wear the wings of the R.A.F. on their right breasts to show that their earlier war service was in that service. A liaison officer from the R.A.F. is also on the strength and, occasionally, a knot of R.A.F. airmen can be seen marching smartly to or from duty. They are familiar with the process of taking Mosquitoes on charge and checking the British equipment with which they are fitted.

High Standard of Serviceability

The station (or group) is commanded by Col. Leon W. Grey, with Major A. E. Podwojski as Group Deputy-Commander. Major Roy Ellis-Brown—one-time flight lieutenant in the R.A.F.—brings his operational experience into use as Group Operations Officer. This post combines the duties of Wing Commander Ops and Wing Commander Intelligence in the R.A.F.

The Americans are quite frank in their praise of the "hot ships" and in their pride they manage to maintain an extremely high standard of serviceability. This is better, per-

haps than that achieved by the R.A.F. I asked them what their major troubles might be, and they assured me that their only bother was to obtain crews good enough for the job. That was how they put it, but what they really meant was that they had some difficulty in getting pilots of the right temperament. But more of that anon.

When they were assessing the number of Mosquito XVIs which would be required for establishment and replace-

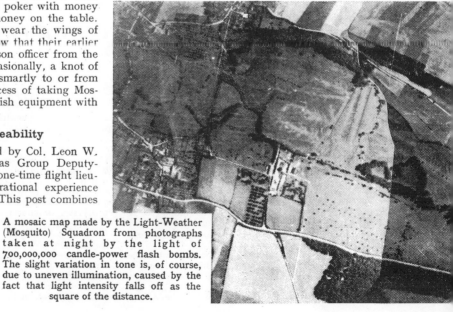

A mosaic map made by the Light-Weather (Mosquito) Squadron from photographs taken at night by the light of 700,000,000 candle-power flash bombs. The slight variation in tone is, of course, due to uneven illumination, caused by the fact that light intensity falls off as the square of the distance.

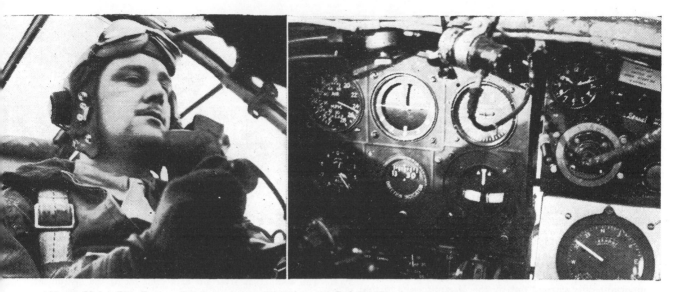

(Above) Major Ellis-Brown, D.F.C., one-time Flt.-Lt. in the R.A.F. The helmet and earphones are of British pattern because the Mosquito inter-com. is of standard type. (Right) Gently cruising on the way home from the Continent. Slightly losing height at 300ft. per min. Present height, 3,400ft. Speed, 250 m.p.h. Direction, 312 deg. Note the slight variation between the gyro direction indicator and the distant-reading compass on the right.

ment, the figure was put on the generous side because they expected that the casualty rate over Germany would be rather high. De Havilland's deliveries were to schedule, and casualties have proved to be lower than expected.

The units which go to make up the station, or group, as the Americans term it, are a Light-Weather Squadron, a Heavy-Weather Squadron, and a Training Unit. It must not be thought that the names heavy weather and light weather refer to the atmospheric conditions in which the squadrons work. It merely means that the one squadron flies Mosquitoes and the other flies Fortresses. The division of labour is roughly this: the Heavy-Weather Squadron does periodic trips out into the Atlantic for a thousand miles and back and at stated intervals along the route observations have to be taken from sea level to 24,000ft. The Fortresses employed on this work have been specially modified.

The Mosquitoes of the Light-Weather Squadron are used wherever the enemy are likely to be encountered or where speed is an essential. Four crews are always standing-by, ready to take off for anywhere from eastern Germany, Czechoslovakia, or Norway, down to the Spanish border.

THE OWNER : Col. Grey, who commands the station watches his ground crew prepare his own pet Mosquito.

or for a quick trip westwards. Usually it means that a weather front is approaching a target area. It is the duty of the crews to find height and base of cloud, direction and speed of front, icing conditions, temperatures and general information. The trip usually entails flying into and through the front to get a check on turbulence.

Navigational Accuracy

One trip was made by Capt. R. L. Lee with Lt. W. A. Biggers as navigator. The job was to test long-range navigation equipment over various areas. They flew from England via Corsica to San Severo, in Italy, where they landed. From San Severo the Mosquito went on experimental trips to Varna on the Black Sea, Belgrade, Budapest and Vienna. At all these points the navigational fixes were within one half-mile to one mile of the actual spot over which the aircraft was flying.

This accurate long-range navigation ties up with another important duty carried out by the Mosquitoes, and that is night photography. In this work it is safe to say that the Americans are well ahead of the rest of the world. As long ago as March, 1929, they were taking experimental air photographs by flashlight over the Capitol and developing and printing the finished pictures on board the aircraft. The small flashlight used then has developed into the 700,000,000 candle-power M.46 flash-bomb used to-day. A light as powerful as this allows of an exposure of 1/25th second This shutter timing is obtained by making a slit travel across the film at a speed synchronised with the forward speed of the aircraft over the ground. So efficient is this arrangement that apertures as small as f/6.3 can be used. The flash lasts about one-tenth of a second, and it is possible, by the employment of photo-electric cells, to ensure that the photograph is taken at the peak of the light intensity.

Twelve flash-bombs can be stowed in the bomb bay of

The wet and dry bulb thermometer mounting on the nose of one of the Fortresses used by the Heavy-Weather Squadron. The weather observer sits at a table facing the instrument.

a Mosquito. The point is, however, that these night photographs, in conjunction with accurate navigation, cover an area sufficient to ensure that the specified target will be included in the picture For many years the Americans used multi-lensed Fairchild T2 and T3 cameras, but a great deal of their war work has been done with the 5in. × 5in. K24, which is the U.S. version of the R.A.F.24 camera.

The Training Unit has three dual-control Mosquito VIs for the pilots to learn to handle their new craft. Seventy per cent. of the Mosquito crews have done 700 hours or more on Fortresses or Liberators before being posted to this much-sought-after assignment. The conversion course includes 5 hours' dual, and another 20 hours' solo flying are put in. Of the pilots who are posted for training, some four out of five make the grade, and the remainder are returned to the pool.

An American Mosquito XVI of the Light-Weather Squadron, seen from under a Fortress of the Heavy-Weather Squadron. Both types are used, the Fortress over the Atlantic and the Mosquito wherever the enemy may be encountered or where high speed is desirable

Great Skill Required

As I pointed out earlier in this article, the Americans are having some difficulty over their crews. Primarily the whole American air set-up is designed to work by day. For this work, however, the majority of the flying is in bad weather by night, and a different standard of pilotage and navigational skill is required. Most of all, there is a different temperament to be acquired. There is a vast difference between being one of a huge formation of tightly packed four-engined bombers, each with more than a dozen guns, and protected by a one-for-one fighter escort, in daylight over Germany, and being in a comparatively small aircraft, and flying in ten-tenths cloud at night to assess the icing conditions. That such a high percentage makes the grade is a great tribute to the basic training and natural aptitude of the U.S. aircrews.

Our own Mosquito XVI crews, flying at high altitudes, have all been attacked from behind or beneath recently by German jet-propelled aircraft when on daylight missions. So it is with the Americans. One such attack was experienced near Stuttgart by Lt. Kenny and his navigator, Lt. Kuehn. The Mosquito was "jumped" by an Me 262 at 30,000ft., and the fight went on all the way down to 12,000ft., at which altitude the Me had run out of ammunition and was leading a futile existence making dummy passes. The Me's guns appeared to be bigger than 20 mm. and the shells were fused to explode at a predetermined

American photographic staff bringing aerial cameras to mount within a Mosquito. Most of the photographs are taken from very great altitudes.

distance. No tracer was used, and there were no hits on the Mosquito. All the way down violent evasive action was kept up, and the pilot estimates that the Mosquito exceeded 700 m.p.h. in the steepest dive. The Me jet fighter was able to keep up.

Speed and Mach Number

The figure of 700 m.p.h. can scarcely be accurate, even for the superbly streamlined Mosquito. Supposing a terminal velocity somewhere in between 0.8 and 0.9 of the speed of sound—and this would be a very high figure—the limiting speed would be in the region of 610 m.p.h. at 15,000ft. Even at sea level, where the speed of sound is 764 m.p.h., the limit would still be below 630 m.p.h.

There is no end to the interest on this station. All the crews are keen to tell of some particular job done on a Mosquito. Another pilot told how his "ship" came home with 40 "shells" in the wings. This is likely to be misleading to the British reader because bullets are all shells to an American, and the term does not necessarily mean the 20 mm. explosive type. Another Americanism likely to be misunderstood in England is that plugs "blow out" instead of misfire

Many of the pilots' reports speak of St. Elmo's fire on the aircraft when flying in cumulo-nimbus cloud. The crews derive a certain Mephistophelean enjoyment from making long sparks come from their finger ends to parts of the machine.

I accepted Maj. Ellis-Brown's invitation to go and have a look at the weather and things in Holland. We made the trip in M for Maypole. The engines, which were already warm, started immediately, and after the double doors of the pressure cabin had been closed by the jockey-capped mechanic, we taxied gently at 1,200 r.p.m. round the perimeter track to the take-off point. On the way the oxygen was connected and turned on and the airfield control told that we should be airborne for a little over 1 hr. 30 mins.

Because of possible overheating, the caravan at the end of the runway was requested to clear a Fortress which was in our way. With the Fort out of the way we were given the "green" and were airborne 24 sec. later, with 140 m.p.h. showing on the A.S.I.

Away on the port bow an ugly black column of smoke rose from just outside another airfield. A Fortress had just crashed and caught fire. Soon after take-off the pilot

A light load. Twelve 700 million candle-power photo flash bombs in the bomb-bay of a Mosquito.

MOSQUITOES OF THE 8th U.S.A.A.F.

The unfamiliar jockey cap sported by American ground crews appears at the entrance. Note the double doors of the pressure cabin. The outer door opens downward and the inner door opens upward.

changed over to the long-range fuel tank in the bomb bay. Altogether some 750 gallons of fuel were carried.

We were in no hurry and climbed leisurely at 1,000ft per min with the Rolls-Royce Merlins doing 2,600 r.p.m. and 210 m.p.h. showing on the "clock." Strangely enough, as we flew over the sea and under an even layer of stratus cloud, we ran into small areas of turbulent air. In such conditions one would expect absolutely bump-free air.

Above was a formation of Fortresses getting height before crossing the Continental coast. Below was a blue-grey carpet of sea and haze which made a perfect camouflage background to the Mosquito. From quite a short distance away it would have been difficult to discern the outline of the aircraft, and the only discernible indication would have been the red outlined star and bar of the American markings. As we approached Holland, Maj. Ellis-Brown pointed away to the left, where a formation of American heavies was getting quite an amount of *flak*.

The journey at this point was not without interest for both of us when I compared the height at which I was flying with the 2,500ft. at which I flew over the same route in a Stirling to Arnhem. Maj. Ellis-Brown was also flying a Stirling when he was in the R.A.F. He went on some of the very early daylight raids, including the attack on the Potez aircraft factory at Meaulte. For him the particular interest in this trip was that he was very close to the part of the Dutch coast where, in 1941, he made three very determined attacks on an oil tanker and succeeded in sinking it. For this he was awarded the D.F.C. He was most generous in his memories of the Stirling. How sturdy it was and how throwable-about! The outward journey over, we turned for base; the pilot asking for a bearing over the V.H.F

How gently the Mosquito seems to fly. It never throbs or drums or seems "busy.' As we returned through the, 11,000ft. level, our true ground speed was not far short of 400 m.p.h., yet a bus doing 20 m.p.h. would make more fuss about it.

Over the coast and down to 1,000ft. Fuel pump off, oxygen off wheels down, flaps down and a perfect landing.

An interesting trip.

FLYING BOATS FREED FROM ICE

SUNDERLAND and Catalina flying boats of R.A.F. Coastal Command which, a few days ago, were frozen in their base in Northern Ireland are again hunting for U-boats in the Atlantic.

When the sea froze over for the first time in fifty years there was ice four or five inches thick, and the temperature fell to 24 degrees below freezing point.

Working day and night for five days in bitter cold and blinding snowstorms, the airmen of the Marine Craft Section hauled as many flying boats as possible on to the shore and broke the ice round the mooring areas of the remaining aircraft.

"Five floating refuellers patrolled for twenty-four hours breaking the ice," said Sqn. Ldr. A. T. B. Cooper, of Ellerby Lodge, Hinderwell, Saltburn. Yorks., who is in charge of the Marine Craft Section—the biggest ever assembled in a lough. "They used bomb scows to clear a space round the aircraft and dragged the ice blocks to the land. They did a grand job."

M. OF S. EXHIBITION

THE Ministry of Supply (Directorate of Public Relations) gave a private exhibition in Manchester recently, extolling the part played by technicians in the current air offensive. The exhibits were well displayed and rich in detail.

The foremost exhibits were a section of a Rolls-Royce "22" Merlin engine and a hydraulically operated gun turret. Of exceptional interest was the device for viewing stereoscopic air photographs.

Three films were shown: a news item, a documentary called the "Sky Giant" depicting the production and use of the Lancaster bomber, and a film showing the production of aircraft engines at a Rolls-Royce factory, emphasising the scrupulous care paid to detail, and the accuracy and quality

Hardly anything connected with the equipment of a modern bomber was left out of the exhibition, which, among other things, included an inflated rescue dinghy, complete with fishing tackle and instruction booklet (how to land a turtle!).

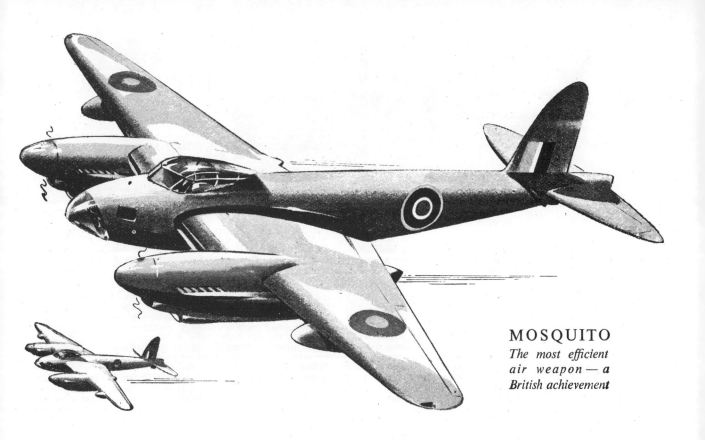

MOSQUITO
The most efficient air weapon — a British achievement

DE HAVILLAND ENTERPRISE

The de Havilland enterprise is a "family" of manufacturing and servicing establishments spanning the world. It has been built up steadily since the last war, and provided a strategically dispersed production system in readiness for the second world emergency. It exists to-day as a unique Empire-wide enterprise renowned for design ability, remarkably equipped to serve aviation in the coming period of reconstruction.

Great Britain - Australia - Canada - India - Africa - New Zealand

Flying the D.H

By Flight Lieutena...

The following article on the de Havillar... Mosquito is by a pilot with experience of flyir... several Marks of the type. It is not expected... nor intended to be of great interest to opera... tional pilots who regularly fly this particula... aeroplane and may have more experience ... it than the author. "The Aeroplane" must...

FIRST IN THE FIELD.—The prototype Mosquito W4050 takes off for a test flight during the original manufacturers' trials. This prototype bomber had two Rolls-Royce Merlin 21 motors and a span of 52 ft. 6 ins.; the short nacelles are also noticeable.

THE FIRST thing that strikes one on looking at a Mosquito is its beauty of line, its clean, smooth taper, and many have remarked that this D.H. product recalls a thoroughbred racehorse.

For a normal-sized man to climb into a Mosquito while wearing a seat-type parachute is almost a physical impossibility (particularly in the case of the Marks III, VI, etc., which have the door in the side instead of in the floor), so the best entry is by climbing the ladder, hauling up the brolly and then putting in some hard work laying it out in the seat. The pilot can then descend the ladder again, take a quick look under the motor nacelles to check that the undercarriage locks have been removed, glance at the top of the fin to see that the pitot tube is uncovered, and do a manual check to see that the rear hatch is securely fastened, for should this hatch blow off in flight, serious things can happen to the empennage.

Once settled in the seat and the harness done up, the pilot is impressed by the comfort of the seat, and the ease of accessibility of the ancillary controls—two points which some constructors might note as being of particular interest to pilots. The system of operating the petrol cocks is also too simple to be readily credited to all constructors of British operational aircraft. Just behind the pilot's seat are two cocks marked " Left " and " Right," and they can be turned so that they are both pointing inwards (towards each other) or both pointing outwards (away from each other); this signifies inwards for inner tanks and outwards for outer tanks. I am writing now of the ordinary production Mosquito which was not fitted with fuselage tank, drop tanks, rocket projectile rails, Radar or any other operational excrescences, and the normal system of petrol-cock operation is, briefly, that you take-off and land on your outer tanks and fly at other times on your inner, or main tanks.

The two Rolls-Royce Merlin motors have to be primed from the outside by the airman on the ground by hand, and have not the advantages of the U.S. practice of electric priming operated by the pilot from inside the cockpit; however, very seldom does a Merlin become temperamental and refuse to start, and so soon as the coolant temperature has reached 60 degrees each motor may be run up, the magnetos tested, and the pitch controls exercised. Since there is an hydraulic pump on each motor, one motor should be throttled right back and the flaps lowered and raised with the other motor set at 2,000 r.p.m., and then the procedure reversed.

For taxi-ing, the aeroplane is responsive to the throttle, and with a bit of practice very little brake will be found necessary. However, when the brakes are used, either for taxi-ing or at the end of the landing-run, they should be used in short, sharp squeezes of the brake lever, as holding the brakes on is inclined to make the aeroplane swing about in a rather untidy manner.

On arrival at the take-off point a brief cockpit drill is carried out as follows. (I think that most people have their own method of checking the cockpit, but the following method is quite satisfactory.) Set the elevator trimmer to one-and-a-quarter degrees nose down and the rudder trim to neutral. Check that the oil temperatures and coolant temperatures are within the laid-down limits, and tighten the throttle friction nut. The mixture control is entirely automatic, so no check is necessary on this. The pitch controls are, of course, in the fully forward position, giving maximum r.p.m. (approximately 3,000 at take-off boost, which is between plus 8¾ lb. and plus 12 lb., according to the mark of motor installed). Check

that the petrol cocks are on the outer tanks, and that the fuel gauges are reading satisfactorily. The radiator shutters, which are worked by the brake-pressure system, are controlled by two small switches usually mounted next to the rudder trimmer, and are either fully open or fully closed (on certain marks they are manually operated by two hand cranks) and for take-off they should be set to fully open. The generator (which is driven by the starboard motor only) should be switched on and the gyros checked. After this the pilot need only clear each motor for a couple of seconds, lower about 15 degrees of flap and he is ready to take-off. On a decent-sized runway there is really no need to use flap for take-off, but the Mosquito takes rather a long time initially to reach her single-motor safety speed after take-off, and in practice the use of a little flap has been found to lower this safety speed in proportion to the time taken to reach it.

While taking-off the throttles should be opened rather more slowly than on most other aircraft, and in general practice to run up the motors and " hold her on the brakes " is entirely unnecessary. As the throttles are opened a Mosquito tends to swing slightly to the left, which is checked by advancing the

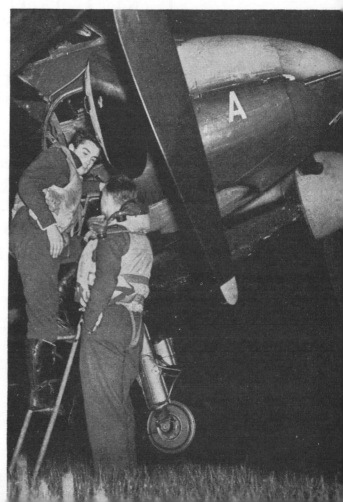

NIGHT TRAIN TO THE CONTINENT.—A pilot and observer about to climb into their Mosquito VI night-fighter in preparation for an intruder sortie over France and Belgium.

MOSQUITO

. P. Hanson-Lester

NIGHT SKY RIDER.—de Havilland Mosquito II night fighters achieved considerable success in thwarting the Luftwaffe attacks on London. Above, a Mosquito II is seen in the matt black camouflage of 1942.

owever, have many readers who have never own Mosquitoes, who are never likely to fly em, perhaps, even, who don't want to fly them, t who will welcome the opportunity of arning something about the handling qualities f one of the outstanding aeroplanes of this Var in both the West and East campaigns.

port throttle a little ahead of the starboard throttle. Until one is thoroughly used to the aeroplane (by which time it should not be necessary) take-off swing should be checked by use of the brakes. If the throttles are opened at all "ham-fistedly," or too rapidly, the swing will develop quite viciously and tend to become almost uncontrollable, a fact which reacts most unfavourably upon the undercarriage, but if they are opened smoothly and slowly the swing is scarcely perceptible, even although the aeroplane is accelerating very rapidly. By the time the motors have been given approximately plus 1 lb. or 2 lb. boost, the tail should be well up and plenty of rudder control available and then the rest of the boost can be poured in. The throttle levers have a stop at take-off boost setting, but by releasing two trip catches on the levers a pilot can go through the gate and obtain approximately another 6 lb. of boost, but the use of this is seldom found to be necessary in general practice.

The aircraft can safely be allowed to leave the ground at about 110-115 m.p.h., and a slight movement of the control column to the right is usually necessary to prevent a tendency to drop the left wing. The undercarriage comes up at quite

a normal rate, and when the lights have gone out (there is, by the way, no visual indicator for the position of the retractable tail-wheel) the pilot can throttle back to plus 6-lb. boost and 2,650 r.p.m. and climb to the desired height. The indicated airspeed during the initial part of the climb should be 150 m.p.h. until the flaps are raised, when it will very rapidly increase to 180 m.p.h., which is 10 m.p.h. above the safety speed for single-motor flying without use of flap. The actual raising of the flaps produces a very strong nose-down effect, but very slight backward movement of the elevator trimmer will correct this, as the trimmer itself is very sensitive.

When the coolant temperatures have dropped below 95 degrees (once again certain marks of motors run at higher temperatures than this) the radiator shutters may be closed, and this action—initially much to the pilot's surprise—also produces strong nose-down effect. These little shutters are set underneath the radiator, which is built into the leading edge of the wing and are similar in appearance to the very small s.t.e. flaps which have been mounted half-way along the wind chord, and they are extraordinarily strong; so much so, that if you trim the aeroplane to fly straight and level with " hands off," and shutters open, then smartly close the shutters, the resultant nose-down effect is so strong that the motors will cut under the effect of negative G, and the aeroplane will go into a steep dive. However, in normal usage the form is to close the shutters with the right hand and retrim the aeroplane at the same time with the left hand.

Having climbed to the desired height, the pilot throttles back to plus 4-lb. boost and 2,400 r.p.m., and changes the petrol cocks over on to main tanks. Naturally, if he is going to climb up pretty high, he should have changed them earlier, but for the sake of argument he is just going to have a little ride around at about 3,000 ft. At the present revs. and boost, which is the optimum for indefinite cruising, the aircraft settles down to an easy cruising speed of approximately 270 m.p.h., and the time has come to start turning gently round the sky in order to feel the controls. The Mosquito is very positive in the pitching plane and extremely sensitive to the trimmer; thus in actual cross-country flying, to trim it to fly hands off and maintain a constant height is not easy; however, only a light touch of the hand resting on the control column is necessary to remedy this. In the rolling plane she is almost unbelievably light on the ailerons and in the yawing plane she does not react favourably to coarse use of the rudder; gentle and medium turns can be very satisfactorily effected with the feet resting on the floor instead of on the rudder pedals without the top needle of the turn and bank indicator registering any protest.

To stall a Mosquito both throttles must be pulled back and the nose raised. With the throttles fully closed and the horn blaring (there is no switch to eliminate this, so it has to be endured as a gentle reminder) the pilot allows air speed to drop off and cannot help observing that she loses speed very slowly in comparison to some of the less streamlined flying Christmas trees, and he also notices that there is little tendency to mush and lose height as he approaches the stalling point. Observation of the actual speed at which the aeroplane stalls is not easy as the needle of the A.S.I. flickers very violently, but plenty of warning is given before the stall because at about 125 m.p.h. the aeroplane starts vibrating and juddering in no uncertain manner. At about 115 m.p.h.

CLEAN ENTRY.—The crew of a Mosquito IV bomber climbing into the cabin through the small door in the floor of the fuselage. When entering the Mosquito care has to be taken to avoid the airscrew when the motors are running.

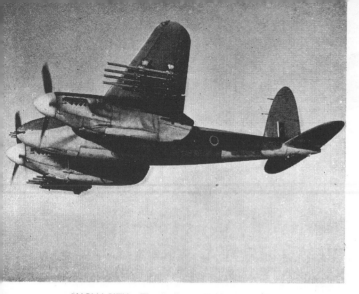

PUGNACITY.—The well-armed Mosquito VI carries, in this photograph, eight rocket-firing projectiles and four .303 machine-guns and two 20 mm. cannons.

she stalls completely, but not in any way viciously, and with flaps and wheels up, the nose drops quite sharply but without any tendency to drop a wing.

With flaps and wheels down and the motors throttled right back there is little difference in behaviour except that the speeds of the preliminary warning and the stall itself are about 15 miles an hour less, the nose has to be held a little higher, and she loses a little more height before actually stalling. To stall with flaps and wheels down and about minus one boost and 2,400 revs. on each motor is not quite such a carefree manoeuvre, but to see how she behaves in this condition is interesting, and does not really present any difficulty. The air speed gets down somewhere in the region of 85 to 90 m.p.h. and the nose has to be held in an uncomfortably high attitude and the aeroplane feels quite reluctant to stall at all! However, a little perseverance coaxes her into the stall, and the nose whips down quite viciously with a tendency to drop the right wing. There is naturally a pretty violent change in the attitude of the aeroplane now, but by applying the normal corrective methods of stick forward, throttles open and picking up the wing with firm (but not necessarily full) opposite rudder, normal flight is regained with remarkably little loss of height. Naturally when the recovery from the stall is effected with flaps lowered care must be taken to see that the air speed does not exceed the maximum allowable for use of flap, which is 150 m.p.h., in the ensuing dive.

So far we have seen that the Mosquito is easy to take off, extremely light and pleasant to handle in normal flight, and, unless treated in a thoroughly unladylike manner, is not unkind in stalled conditions. Let us now see how she will react in the event of one of her " fans " going off duty.

We will assume the pilot is flying peacefully around the sky at 3,000 ft. at 270 m.p.h. with plus four boost and 2,400 revs. on each motor and, since the generator is driven by the starboard motor, the port airscrew is feathered. This is very rapidly done by pressing the port feathering button on the dashboard and holding it in for a second just to make sure

DOUBLE FACED.—The Mosquito VI could be used as a fighter-bomber and, in this photograph, carries two 500 lb. bombs as well as machine-guns.

that the solenoid has engaged. Immediately the revs. start to decrease, and as they do so the pilot fully closes the port throttle and in a very few seconds the port airscrew is fully feathered and stationary beside. The aeroplane's reaction to this is a fairly sharp swing to port, but, owing to the servo rudder control, this is very easily corrected with a gentle but firm pressure of the right foot on the rudder pedal. A fair amount of rudder trim (about two or three turns) relieves this, and, having opened the starboard radiator shutter, slight application of tail-heavy trim on the elevator, enables the aeroplane to be flown " hands off " with no trouble. With normal load at this height there is no necessity to open up the starboard motor from its present setting in order to maintain height at approximately 180 m.p.h., but at 2,650 revs. and plus six boost an easy cruising speed of 200 to 210 m.p.h. can be maintained. The aeroplane can also be trimmed to execute with " hands off " perfect rate two turns without loss of height or airspeed in either direction! This is indeed " a little something which the others haven't got." And when the controls are taken in hand again, and the starboard motor opened up a little further, quite tight, steep turns in either direction can be done. I have even watched Geoffrey de Havilland, junior, carry out upward rolls with one motor feathered, although I must confess I have not aspired to try it myself.

While on the subject of aerobatics I do not propose to go into the details of how to perform each individual manoeuvre, but will only say that normal aerobatics present no difficulty whatsoever, manoeuvres in the rolling plane being particularly easy to execute. Nor do I intend to go into the operational aspect of the aeroplane or discuss its maximum speed under various conditions or its climb and performance figures at various altitudes. Almost all normal aerobatics can be carried out with no trouble at all (but the " Mossy " is not

PEACEFUL GUISE.—The Mosquito XXX has been adapted for civilian use and here is a Mark III with its civil markings and British colours.

a good aeroplane in which to practise either right- or left-hand spins!) but if you want to push the nose down and throw her about a bit, you will not find her short of knots.

However, to return to more peaceful pursuits; we will now assume the pilot unfeathers the port " fan " and see how he gets on with a normal approach and landing. Before we leave the subject of the Mosquito's amazing single-motor performance I would like to add that the single-motor approach and landing is easier to do well, either with or without an airscrew feathered, than on any other type of twin motor aircraft which I have flown, and I have on more than one occasion started an overshoot procedure at 500 ft. and gone round again with flaps and wheels down without any difficulty whatsoever, provided always that a safety speed of at least 145 m.p.h. has been maintained on the approach. I must also admit that I have seen people endeavour to go round again after their speed had dropped below this figure, and on all occasions they came to grief.

To unfeather the airscrew the pilot first of all checks that the port ignition switches are on and that the port pitch control is in the fully coarse position. He then opens the port throttle to the normal starting-up position (about ¾ in. open), pushes in the port feathering button and holds in it. If he looks at the port airscrew he can see the blades gradually twisting to assume some positive pitch and in a second or two the airscrew starts to revolve. Almost immediately the motor fires and the revs. start to build up on the revolution indicator, and when they have reached 1,000 r.p.m. he releases the feathering button and slips one finger into the little hole behind it to make sure that it has returned to the normal position. This is rather important, because just occasionally the solenoid sticks and the button does not come out, with the result that the revs. are pushed up by the unfeathering mechanism until they are beyond the control of the constant-speed unit, and once the airscrew starts to windmill freely the revs. build up to about 3,750 r.p.m., and the poor old Merlin starts to make a noise like a mass formation of Harvards taking-off, a state of affairs which is not recommended in the Rolls-Royce Merlin handling notes. (The cure for this eventuality is to pull the button out by hand and raise the nose fairly steeply; this produces a drop in airspeed, and since the airscrew is not governed by a c.s.u. the revs. drop back until they eventually come within the range of the pitch control and are once more governed by its manual setting in the cockpit.)

Now the motor is turning over at 1,000 r.p.m. (although in practice they may have built up a little higher), and if the coolant temperature is such that to warm up at low revs. is not necessary the pilot can move the constant-speed lever forward until both motors are showing the same r.p.m.; he then moves the throttle forward until the boost readings for both motors are the same, at the same time neutralizing the rudder trimmer. If he now closes the starboard radiator shutter he returns to our previous happy state of cruising at something over 250 m.p.h.

On arriving back at the aerodrome and having selected the runway, the pilot descends to about 1,500 ft. and throttles back as far as he can without being inflicted with the horn blowing. He is then endeavouring to lose enough speed to permit him to operate the landing gear, and at this stage of our flight the excellency of the Mosquito's streamline is most apparent. To lose the hundred odd miles per hour which is the difference between cruising speed and the maximum permis-

FEATHERED BUT NOT FRIGHTENED.—Mosquitoes can fly and climb on a single motor, and here a Mark IV is maintaining level flight with one airscrew feathered.

setting to give the desired rate of descent to the rapidly approaching runway beneath. As he turns into wind the air-speed must not be allowed to drop below 140 m.p.h., as in the event of a motor failure he wants to be able to have at least 145 m.p.h. available in order to maintain perfect control. Thus he maintains this seemingly high-approach speed right down until coming over the hedge. At about 600 ft. the pilot eases the pitch controls to the fully fine position and settles down to the final approach. As the aeroplane is crossing the hedge the throttles are cut, the speed drops off to about 125 m.p.h., and the aeroplane is eased on to the ground. I think that the Mosquito is one of the easiest aircraft there is in which to make a good landing. whether it be a tail-high " wheeler " or in a three-point attitude. As the aeroplane is checked close to the ground she loses speed rapidly, and by moving the control column back a little to prevent her sinking too rapidly, she just cushions on to the ground. Without using much brake, the " Mossie " has a very short landing run in relation to such a high approach speed, and with a 10 to 15 m.p.h. wind nobody should have any difficulty in landing in 600 yards; it has been done many time in a 450-yard field.

If a really smooth landing has been made and there is no appreciable cross-wind the " Mossie " has no tendency to swing at all, and will just quietly roll to a stop, but if the wind is a little on the port side, or if the arrival is not quite in the highest traditions of the Central Flying School, she may tend to swing slightly to port. This is easily corrected by a short, sharp squeeze of the brake lever. (May I remind you here that prolonged holding of the brakes nearly always result in over-correction? Two or three short " puffs " will cure any swing which may occur under normal circumstances.)

All that is left now is for the pilot to raise the flaps, taxi back, disperse her, turn all the taps off, and the flight is over. My own opinion is that there is no aeroplane more delightful to fly in than the Mosquito, and should readers who have more recent practice on them than I disagree with my procedure, I can only reiterate that this article is intended purely as of objective interest to those who have not driven one.

FLIGHT TEST.—The Mark VI Mosquito fighter-bomber in this photograph is receiving a final ground check before being tested in the air. The housing for the lower guns can be seen opened up.

sible speed for lowering the wheels and at the same time maintain a constant height, takes much longer than one would expect. On the other hand, when the pilot has lowered the wheels at 180 m.p.h., he has to put on as much as plus 2 lb. or 3 lb. boost in order to maintain 160 m.p.h. without losing height. This is because the somewhat massive undercarriage destroys the smoothness of the Mosquito's form and she becomes quite a different aeroplane to handle, losing much of her sensitivity.

At this stage of events the petrol cocks are turned over to outer tanks, the radiator shutters opened, and the aeroplane trimmed to continue flying at 160 m.p.h.. As the pilot turns across wind, he lowers the flaps; these can be let down all at once, and have a very powerful nose-up effect. This is naturally countered by easing the nose down and winding the eleva-tor trimmer well forward, trimming the aeroplane to fly at 140 m.p.h. with flaps fully down, and adjusting the boost

MOSQUITO OFFICE.—The dashboard of the Mosquito in the pilot's cockpit compares very favourably with other aircraft cockpits for simplicity of layout.

Bank Raids

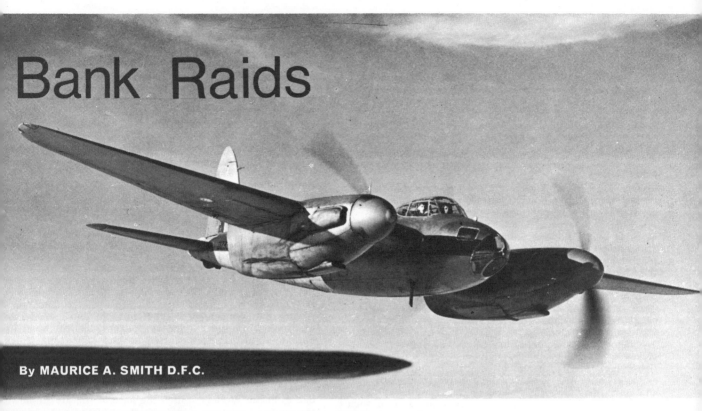

By MAURICE A. SMITH D.F.C.

The part played by Mosquitoes in bursting canals

"**S**O YOU'RE IN THE Fifth Air Force", they'd say with a touch of sarcasm and of envy. True, 5 Group often did operate as something of an independent air force, and there was a suggestion of the elite. It also had its own Pathfinders, closely related to those of 8 Group who started the idea, but operationally separate and using some different techniques.

The broad background to Pathfinding is well known. Bomber squadrons started the war unable to find a target at night, let alone hit it, and many aircraft were lost trying in vain. After a while some experienced crews began to do slightly better, and it occurred to those in command that a few outstanding ones might be able to lead the others. But even the best crews needed a lot more help in finding any but easy targets. Soon, improved navigational aids, including radar, began to emerge. Means of marking target areas at night were evolved.

Even greater precision was now called for, and one answer came from 5 Group, which was ordered to learn how to hit pin-point targets, including those in France, Holland and other friendly but occupied countries, where stray bombs were not just wasted, but might kill our allies. Limited numbers of Lancasters had to be able to identify and hit small high-priority targets anywhere in German-held Europe.

It became clear that, to do this with a fair degree of certainty, not just the target area but a precise aiming point had to be found and illuminated by experienced crews with special equipment. A few low-level markers in Mosquitoes then had to go in and put target indicators right on the spot selected. Finally, experienced crews of the main force of bombers would come over and aim at these accurate target indicators which took the form of tight clusters of coloured incendiaries—usually red.

With all this going on in the target area, it seemed sensible to have someone directing the operations. A master of ceremonies was called for. So low-level Master Bombers, Controllers, Masters of Ceremony, call them what you will, emerged as they had done in a slightly different role in 8 Group. They had in common enough experience to know what they were about; speaking voices that came over on VHF clearly, and maybe a little more maturity at around 30 years of age than the majority of bomber crews in their early 20s. Presumably they had been noticed in their squadrons and seen to lead reliable crews. These they very reluctantly had to leave behind, except for the navigator, when the smaller and more agile Mosquitoes came into use.

All sorts of safeguards and refinements in detail were

Left, Dortmund Ems at Ladbergen after attacks by Bomber Command Lancasters. The narrow island between the channels was marked by red target indicators. The banks of both channels were severely damaged and the canal emptied. The River Glane, crossed by camouflage strips, passes underneath and the canal bed above it has been holed. Right, a contemporary November photograph of the crew concerned, with navigator Lee Page on the left. The B.XX Mosquito in the background, with Packard-Merlin engines driving paddle-bladed propellers, carried Loran and Gee. Unlike the B.IV in the heading picture it had no shrouds for its groups of five exhaust stubs, but did have intake guards and 100gal drop tanks.

introduced for the target marking, and failures became exceptional. One outstanding success on November 4-5, 1944, and one failure on the next night of November 6-7, are picked out here for description. Both were concerned with German canals, upon which there were six attacks by 5 Group in the two-month period starting November 4, 1944. There had been one previous successful attack on September 23, which had included 617 Squadron carrying 12,000lb "Tallboys".

After D-day

Germany's communications were under almost continuous attack after D-day. Railways and their marshalling yards had little respite; Baltic ports and shipping were hit; the sea approaches and river mouths were mined—in all of which Bomber Command was very active. These attacks were both tactical and strategic. In the co-ordinated plan of late 1944, which had the specific purpose of ending the war as quickly as possible, zones were defined for attention and, in addition, five specific interdiction targets. Two of these were main-line railway viaducts and three were sections of canals. Then, as now, a chain of great canals formed a most important part of Germanys' industrial transport system.

Some stretches of these huge canals run above the level of the countryside. Rivers and even roads pass under them. It was not too difficult a task to pulverise their embankments with loads of high explosive bombs and so empty the canals for several miles, stranding the long barges and incidentally causing local flooding. Two such vulnerable raised stretches selected for attack were on the Dortmund Ems Canal near Ladbergen (between Munster and Rheine), where it briefly divides into two channels, and on the Mittelland Canal near Gravenhorst, close to its junction with the Dortmund Ems. These are in the area just to the north of the Rhur, and the two canals connect this huge industrial area with Hannover, Brunswick, and Berlin to the east.

For us—for Navigator Lee Page, me and Mosquito B. Mk. XX, KB401 "Easy", that night of November 4-5, 1944, with the Dortmund Ems Canal as our target, started badly. Our morale was always low before take-off, mainly because we operated alone and still missed the bustle and chatter of a squadron departure and the close presence of our Lanc crew of seven. Controllers of 5 Group flew from Coningsby, which was 54 Base HQ and also housed 83 and 97 squadrons. The Mosquito Marker force of 627 Squadron flew from nearby Woodhall Spa. The Lanc squadrons trundled off an hour or more before our solitary Mossie (we pronounced the "ss" as "z's") was due to get airborne. We would start up on the silent tarmac, using only a flash lamp to see by, check time and taxi out for our solitary take-off. At least flying control would tell the ops room we had gone.

"Easy" had been over to 627 Squadron for service and had flown back that morning. She started well and the Packard-built Rolls-Royce Merlins ran up OK. This was only a three hour sortie, so our 100-gal auxiliary tanks under the wings were only half full. Our load was one yellow and two red target indicators—red for marking, yellow to cancel any stray red, or German "spoof" marking.

We lined up on the string of tiny runway lights, opened up against the brakes and started our run. Differential throttle, then rudder to check swing as the tail came up, and, at 105kt, off into the night. Almost immediately the engine notes sounded wrong and the aircraft yawed as we climbed. Check both throttles and bang pitch levers forward on their stops. Wheels up, climbing at 160 kt, height 500ft, straight ahead. The starboard engine seemed to roar and the aircraft tried to slew to port. Perhaps the port

engine had cut? No, all seemed normal there; instruments, noise, exhaust flames long and lilac blue as they should be. Again the roar and lurch, and this time the starboard rev counter sailed up somewhere near the 3,500 mark. We had a prop trying to run away.

In a few seconds of surprise and diagnosis we reached 800ft, levelled out and throttled right back on the starboard engine. There was no question of pressing on, so we must land back. I decided not to feather the starboard airscrew. The engine was unlikely to come to more harm and could help in emergency. A Mossie would not go round again on one engine from ground level with flaps and wheels down, so we would be committed at around 400ft on the approach. Thought: "When did I last practice a single-engine landing at night?"

Now 627 Squadron would have a standby aircraft ready to go at Woodhall Spa. We did not have one at Coningsby. Marker Leader would be quite capable of conrolling the operation if we did not show up—so long as he got there. No, lets have a go. Break radio silence and try to sound like a training flight. Tell Woodhall you want to land and take up a spare aircraft. They will guess what is going on.

They did. We landed without incident and leaped into a strange Mosquito IV, DZ418 "Baker", looked round its unfamiliar instruments, started up and taxied out. Quick engine check and away up to 10,000 ft on course. We flew at 2,850rpm + 9lb/sq in, cut corners on our dog-leg course in dark cloudless sky, and were near enough to see the first parachute flares fall, some minutes before the attack was due to begin. We had stuffed the nose of the Mossie down for the last ten minutes, holding about 360 kt and arriving

War Office, Europe (air) 1/500,000 map of 1945 showing the Münster/Rheine/Osnabrück triangle containing the Dortmund Ems and Mittelland Canals junction. The raised and therefore vulnerable Ladberg and Gravenhurst sections are arrowed, with the Mittelland on the left.

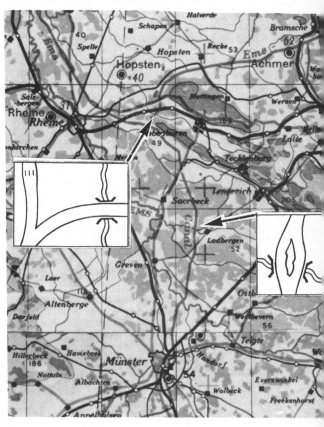

in the Ladbergen area at 1,200ft. Squadron Leader Churcher had got things started. Visibility was good. Using their usual divebombing techniques the Mossies now put their red T.I.s on the aiming point.

The VHF talk would have gone something like this: At zero hour minus ten minutes.
"Controller to Marker Leader, how do you hear me? Over".
No reply. Probably out of range. Zero minus nine minutes call repeated: -
"M.L. to C. clear threes".
i.e. Plainly heard but rather distant.
"C. to M.L. I am running two minutes late. Can see the the flares. Please start the marking".
"M.L. to C. Roger".
"Marker 4, Tally Ho" (meaning that number 4 of the six Mosquito low-level markers has found the marking point and is diving in to put a red T I. on the spot).
"M.L. to Marker Force, that T.I. is 200 yards N.E. of the point, back it up south" (M.L. has decided to accept an error, when backed up south, of about 100 yds.).
"C. to M.L. I am with you".
"M.L. to C. Roger, handing over".
"C. to M.F., back up the red T.I. to south'.
"Marker 3, Tally Ho" "Marker 2, Tally.Ho".
"C. to M.F., thank you that is enough. Clear the area."
"C to Codework Force, come in and aim at group of Red T.I.'s as planned. Codeword Force, aim at red T.I.'s as planned."

A great gaggle of 170 Main Force Lancs dropped their full loads of 14 x 1,000 lb G.P. bombs, many with long delay fuses to add to the confusion on the ground. Some early loads fell too far east, but the error was noticed and corrected. The flares had blown away and long gone out, and even the target marking was practically obliterated.

The attack had been short and concentrated; 930 tons of bombs fell, all but 5% dropped in a circle of 530 yards radius about the MPI. This worked out at about 25 x 1000lb bombs per acre, the best concentration achieved by the group to date. Tomorrow's P.R.U. photos would show the extent of damage and success. Three aircraft had been lost, two to fighters after leaving the target area; and two enemy fighters were claimed shot down.

The Mittelland

Two nights later, we were briefed for a similar attack on the Mittelland Canal at Gravenhorst, only about 15 miles north of Ladbergen. We still hadn't got our Mosquito "Easy" back, so were using a borrowed Mk IV, DZ518, "Mike". The weather was to be similar, but with more haze and a stronger wind. The actual marking point was not very . clearly defined, and we studied the area and the buildings, memorising any features—railway line, salt pans that might stand out. This time we flew out more sedately to flight plan, with a little more time in hand.

As we turned into our last leg for run-in, we saw occasional searchlight beams fanning around to starboard. Quite suddenly, a ruddy great violet-white beam opened up bang onto us, followed at once by others. We had heard about so-called master beams, so perhaps this was one of them. But there should not have been so many lights in the gap between Rheine and Münster.

We were at an effective height for predicted flak, which would probably follow. And it did—but we had moved over, luckily without the searchlights following. The time between guns firing and their shells bursting worked out as a count of one second per thousand feet. (Experts might disagree for the "88s" with their high muzzle velocity.) At 12,000 feet about 12 seconds from gun flash. If you dived 30 degrees to port and looked high to starboard, there

Empty channels and stranded barges in the Mittelland Canal which was missed once but hit twice.

was a good chance of seeing the flak burst where you would have been had you kept straight on. We were glad to be in a manoeuvrable Mossie rather than a Lanc or, Lord help us, a Stirling.

A solitary aircraft was always more vulnerable, and this was one big worry for single Mosquitoes and even more for a main force bomber that got winged and became a straggler without even "window' protection. The first bursts had missed us and we could quickly jink out of range. But why this nasty reception over what should have been open country? Voice from navigator,
"Should be about 12 minutes to go, Skip. I reckon that was Münster. We are a bit off course and ahead of time."
Too right, and our first hint of a met error on wind strength.

For a final, accurate positioning beyond Gee range the Mosquito force in particular, with their more limited space and navigational equipment, depended on a few selected Pathfinder Lancs, with all the blind aids including H_2S radar, to drop green proximity indicators

5 Group attacks on German canals 1944-45

| Date | Canal Target | Aircraft (Lancasters and Mosquitoes) | | | | Tonnage of bombs dropped |
| | | Despatched | | Attacked | Missing | |
		Total	Mossies			
4/5 Nov. 1944*	Dortmund-Ems (Ladbergen)	176	8	170	3	930·4
21/22 Nov. 1944	,, ,,	128	5	128	0	629·2
1 Jan. 1945 (day)	,, ,,	104	2	94	1	558·6
		408		392	4	2,118·2
6/7 Nov. 1944*	Mittelland (Gravenhorst)	235	7	30	10	66·7 (abortive)
21/22 Nov. 1944	,, ,,	143	6	129	2	613·3
1/2 Jan. 1945	,, ,,	157	5	152	2	715·7
		535		311	14	1,395·7

* Described here. N.B. One previous successful night attack on Dortmund-Ems Canal, Sept. 23, 1944.

as near the aiming point as possible. Similar guidance was sometimes given for track turning points. We waited hopefully as we dived gently in the approximate direction of the target and identified with the Marker Leader. No greens. No parachute flares yet.

There was an amber tinged haze in a dark sky and nothing could be seen below. At long last, a few flares showed ahead and left of us, and beyond, a glimpse of a green T.I. A quick caustic word to the navigator and dive hard over towards them. Faint reflections showed here and there on the ground which could have been water or glass. As we dived lower we saw even less. A few flares backed up the dying light of the first ones. Anxious call to Marker Leader. Only about four minutes to go. He was searching. A few more scattered flares were dropping, but at 800ft there was not a single recognisable feature below. The German blackout was faultless. I had a frightful mental picture of the main force bearing down like a great herd of elephants and even more difficult to stop. Search harder.

Tip the Mossie on edge. Marker 4 says he has found the salt pans but practically no flares there to help. Call Flare Force Lancs and Backers-up for anything they've got left—knowing most, if not all, have already flown over. *"Marker 3, Tally Ho!"*
Thank god for that Marker. Where is he? Marker 2 calls for more flares. Just time to save the operation. Call up Main Force to delay their arrival by two minutes. Dog-leg or orbit. I could just imagine what those crews were calling us. Marker 2 is not able to back up. No longer sees the solitary T.I. Marker 3 was sure he had found the spot, but the T.I. must have gone into the canal and been extinguished. No flares left. Four minutes after H. hour. Despairing effort; full throttle climb to 2,000 ft for a more vertical view and maybe a glimpse of a silver thread of canal. No hope. Too late. Attack should have been over by now. Dangerous to keep the main force milling around any more. Half Germany must know we are blundering around the sky north of the Ruhr. *"Abortive code sign to Codeword Force: Go home, Go home."*
There was no alternative target. In fact 30 of the Lancs, including markers, got their loads away, presumably blind, using their own H$_2$S. Most of the bombs had delay fuses, so we did not see them explode. Marker and Flare Force Lancs usually filled up any spare stations in their bomb-bays with 1,000 pounders.

A frightful nightmarish experience it was. What on earth could one say to the AOC—who expected a direct, personal call and first assessment, day or night, as soon as the Controller landed back and found a scrambler phone in the debriefing room?

A raid assessment showed that an exceptional combination of troubles had made this attack abortive. A much stronger wind than expected blew such parachute flares as there were away to the east and scattered them. No fewer than seven of the 14 Flare Force plus 4 Blind Marker Lancaster crews had suffered H$_2$S radar failures. The one Target Indicator that Mosquito Marker 3 got down in about the right place fell into the water and could not be seen or backed up by others. The green proximity markers (two in this case), which were sometimes accurate enough to serve as emergency aiming points, could not be assessed and were confused with a green route marker. Anything other than accurate bombing on such a target, out in open country, would have been useless.

Unhappily, ten of the 235 aircraft were lost, for reasons not altogether clear. The relatively bright sky and good upper visibility helped enemy fighters in the target area and on the way back. The confusion and concentration of aircraft around the target would have been conducive to collisions. Landing back with a full bomb load was no-one's idea of a picnic . . .

Of course, the Group went back to Mittelland Canal a couple of weeks later, and, with only slight opposition, found and emptied it properly. Then, a month later, when frantic work on the part of the Germans had just finished patching it up for use again, it took an even bigger pounding, this time in daylight. At the same time, the Dortmund Ems was also successfully attacked and emptied. Traffic through Münster Locks was reduced from a 1944 monthly average of 844,000 tons to N and E and 585,000 tons from N and E to 14,000 tons and 11,000 tons respectively in January, 1945.

Extracts from official reports speak of 5 Group's second attack on Ladbergen. It "breached the canal at the same place, but this time the breach on the western by-pass was much wider. The eastern by-pass was also hit and two lengths of the embankment totalling about 1,500ft were destroyed. Two bombs pierced the viaduct over the River Glane . . . the water, carrying barges with it, flowing into the surrounding countryside . . . the Mittelland Canal was also drained for the first 18 miles because a safety gate had been left open".

Bomber squadrons involved in those November operations, all flying Lancs, were 44, 49, 50, 57, 61, 106, 189, 207, 227, 463, 467, 619, 630. Those concerned with the marking were 83, 97, and 627. To quote from the *"Strategic Air Offensive against Germany 1939-1945"* Volume 111, on the programme of canal bursting raids, "It may be doubted whether the accuracy, regularity and effectiveness of those brilliant operations had ever, in combination, been approached by any air force in the previous history of bombing".

Grateful acknowledgement is made to the Air Historical Branch of the Ministry of Defence, the Public Records Office and the Imperial War Museum, for assistance in the preparation of this story.

preservation profile

D.H.98 Mosquito T Mk III G-ASKH/RR299

Preserved by British Aerospace at Hawarden, Cheshire

Above, *flying with the Home Command Examining Unit in October 1958.*

Above, *landing at Exeter in 1962 returning from a CA-ACU flight.*

Above, *RR299 as she appears today, in 633 Squadron markings.*

Above, *at a Farnborough airshow in the mid-Sixties.*

The story of the birth of the remarkable de Havilland Mosquito is now aviation folklore and need not be repeated here. Suffice it to say that de Havilland's chief designer R. E. Bishop and his team designed the D.H.98 as a high-speed, unarmed bomber, made of wood and powered by two Rolls-Royce Merlin engines. The Air Ministry could see no future in an all-wood bomber and so de Havillands proceeded on their own. The design staff moved to Salisbury Hall, an historic house situated just up the road from Hatfield, so that work could be carried out in secrecy and without undue interruption. The full story of W4050, the prototype Mosquito, was told in the *Preservation Profile* in our April 1974 issue.

The amazing performance of the prototype Mosquito resulted in huge numbers of production aircraft, in a variety of roles, so that by the time the Mosquito TT.39 – the ultimate derivative – emerged, a total of 7,781 Mosquitoes had been built. Production lasted nearly ten years and the last RAF Mosquitoes were retired from front line service in 1955, though trainers continued in the Civil Anti-Aircraft Co-operational role until the early Sixties.

The first two-seat dual-control trainer variant of the Mosquito, the T.III, first flew on January 30, 1942, with deliveries beginning the following year. A total of 350 T.IIIs was built, mainly at Leavesden in Hertfordshire. One of these aircraft, RR299, is the subject of this profile.

Mosquito T.III RR299 was built at Leavesden aerodrome early in 1945, rolling off the production line in the first week of April. RR299 and its sister aircraft '98 were delivered to 51 Operational Training Unit at Cranfield, with RR299 being delivered there on April 14, three days after '98. On June 5, RR299 moved to 27 Maintenance Unit at RAF Shawbury whence it was allocated for duties in the Middle East. After first being flown to No 1 Ferry Unit at RAF Pershore, the Mosquito left the UK bound for Aden, via Cairo, on December 5. A year later it returned to RAF Cosford, on December 31, 1946.

In the following year RR299 was put into storage at 9 MU Cosford until 1949 when, on May 31, it was issued to 204 Advanced Training School at RAF Driffield, where it

was coded FMO-B. Later that year, on December 19, an accident necessitated sending the aircraft to Sywell for repair. In August 1950 RR299 arrived at 22 MU Silloth, and passed to the Ferry Training Unit at Benson in September 1954 for four months before going back into storage, first to 48 MU Hawarden, then to 12 MU at Kirkbride, and finally back to Hawarden in March 1957. From May that year RR299 flew with the Home Command Examining Unit at White Waltham. From February 1959 it served as a hack with Headquarters Fighter Command for a couple of months. From April 1959 RR299 was based at Exeter, with Nos 3/4 Civilian Anti-Aircraft Co-operation Unit, flying as a live target for guns, bearing the code letter X.

Finally, in March 1963, RR299 was retired to 27 MU and in July that year was acquired by Hawker Siddeley. It received a Certificate of Airworthiness on September 10, 1965, and took up the civil marking G-ASKH.

Naturally RR299 was subsequently in much demand for film work. Her two best known roles were in *Mosquito Story* and *633 Squadron*. For the latter film RR299 lost her overall silver scheme for camouflage and the spurious squadron markings HT-E, which are still carried.

Though the Mosquito has flown only 1,300hr in nearly 40 years, shortage of spares means that it has to be flown sparingly.

Thruxton's Mossies

RICHARD RIDING outlines the fates of six D.H. Mosquitoes that appeared on the British civil register in 195

During the early summer of 1956 the peace of the Hampshire countryside surrounding Thruxton aerodrome, near Andover, was shattered by the familiar sound of Rolls-Royce Merlin engines, heralding the arrival of six de Havilland Mosquito PR.XVIs. Bearing temporary British civil registrations, these ex-Royal Navy Mosquitoes, minus their hooks, had been ferried from Lossiemouth Royal Naval Air Station following their sale to R. A. Short, to whom they were registered

Heading photograph *shows Mosquitoes G-AOCK and G-AOCI down on their uppers shortly before being burnt. 'OCK has been retouched to appear as 'OOK.* **Right,** *G-AOCI shortly after arrival at Thruxton, complete with Royal Navy serial NS639, which had been removed by the time the* **bottom** *photo had been taken*

in May 1956. A seventh aircraft, the Mk TR.33 Sea Mosquito registered G-AOCO, was never delivered and remained at Lossiemouth. Initially, registrations G-AOCI to G-AOCY in-

clusive had been reserved fo Mosquitoes in Short's name, but th last ten were not taken up and th marks were allocated to other ai: craft. The six Mosquitoes flown int

D Conway

Above, *this Mk 16, G-AOCJ, went to the Israeli Air Force in 1957 as 4XFDG-91.* **Right,** *two photograpgs of G-AOCK, which was eventually burned at Thruxton in October 1960.*

Thruxton, G-AOCI to G-AOCN inclusive, were parked out in the open, and three were fated to become familiar landmarks for several years.

The other three, G-AOCJ, G-AOCM and G-AOCN, were destined for foreign parts. In August they left Thruxton for Hurn, where they were completely overhauled by Independent Air Travel Ltd for sale to the Israeli Air Force. The first to leave for Israel was G-AOCM, which flew off on October 12 with the marks 4XFDG-90, followed 18 days later by G-AOCJ, re-marked as 4XFDG-91. Finally, on December 1, G-AOCN left for Israel fitted with underwing drop tanks and bearing the serial number 4XFDL-92.

By that time the Israeli Air Force had taken delivery of 300 Mosquitoes from England and France, and these were now in service as fighter-bombers and reconnaissance aircraft. Many Mks 4, 6 and 36 were bought from the French government as scrap for US$200 each. Several years earlier a handful of Mosquitoes had left England under mysterious circumstances, bearing civil registrations, to reappear in Israel. At the time there was an arms embargo on that country.

One of the remaining aircraft at

Mosquito Mk 16 G-AOCL was not civilianised, but was left to rot at Thruxton before being burned in October 1960. During its service with the Royal Navy it had been fitted with four-bladed propellers.

A. J. Jackson

This Mosquito, G-AOCN, was ferried to Hurn from Thruxton in August 1956 and sold to the Israeli Air Force as 4XFDL-92 the following year.

Thruxton, G-AOCI, had experienced a very exciting encounter during its wartime RAF service. As NS639 it was flying in the vicinity of Munich on September 16, 1944, in the hands of Flt Lt Dodd and Flt Sgt Hill, when it was suddenly jumped by two Me 262s. These aircraft made eight separate attacks on the Mosquito before it took refuge in cloud cover at 6,000ft. Both G-AOCJ and G-AOCK had served with the US 8th Air Force.

The Mosquito PR.XVI was a photo-reconnaissance version of the Mk XVI

bomber. Powered by a pair of two-stage supercharged Merlins, its all-up weight was 23,630lb. A total of 432 PR.XVIs was built, of which 24 went to the Royal Navy.

The three remaining Mosquitoes at Thruxton laanguished in the open, gradually deteriorating, until they were dragged from the undergrowth and burnt in October 1960. Only a few

years later, film companies were scouring the country for Mossies to star in their productions.

Thruxton Mosquitoes

G-AOCI	ex NS639 not civilianised
G-OACJ	ex NS742 to Israeli A.F. as 4XFDG-9
G-OACK	ex NS753 not civilianised
G-AOCL	ex RG173 not civilianised
G-AOCM	ex RG174 to Israeli A.F. as 4XFDG-9
G-AOCN	ex TA614 to Israeli A.F. as 4XFDL-9

A. J. Jackson

Right, *Mosquito G-AOCM at Thruxton before flying to Hurn in August 1956 for conversion for the Israeli Air Force. The photograph* **below,** *taken at Hurn the same year, shows the same aircraft in Israeli markings as 4XFDG-90.*

Colin Bruce

A BROOKLANDS AIRCRAFT PORTFOLIO

Hawker
HURRICANE
PORTFOLIO

One of a series comprising technical descriptions—cutaway drawings—genealogy—combat and operational reports from contemporary articles from Flight, The Aeroplane and Aircraft Production, with modern material from AEROPLANE

Hawker

HURRICANE

PORTFOLIO

A BROOKLANDS AIRCRAFT PORTFOLIO

Boeing B-17 and B-29
FORTRESS and SUPERFORTRESS
PORTFOLIO